INSIGHTS *and* PERSPECTIVES

Fifty-Seven Thoughtful Essays

VERN WESTFALL

INSIGHTS AND PERSPECTIVES
FIFTY-SEVEN THOUGHTFUL ESSAYS

iUniverse books may be ordered through booksellers or by contacting:

iUniverse
1663 Liberty Drive
Bloomington, IN 47403
www.iuniverse.com
1-800-Authors (1-800-288-4677)

Because of the dynamic nature of the Internet, any web addresses or links contained in this book may have changed since publication and may no longer be valid. The views expressed in this work are solely those of the author and do not necessarily reflect the views of the publisher, and the publisher hereby disclaims any responsibility for them.

Any people depicted in stock imagery provided by Thinkstock are models, and such images are being used for illustrative purposes only. Certain stock imagery © Thinkstock.

ISBN: 978-1-5320-1587-8 (sc)
ISBN: 978-1-5320-1589-2 (hc)
ISBN: 978-1-5320-1588-5 (e)

Library of Congress Control Number: 2017901149

Print information available on the last page.

iUniverse rev. date: 01/27/2017

FIFTY-SEVEN THOUGHTFUL ESSAYS
By Vern A. Westfall

Essays range from philosophical speculations to short narratives on science, religion, and politics. The essays raise important questions regarding social responsibilities, scientific theories, experimental evidence, and ongoing explorations.

NUMBERED ESSAYS
TITLES AND DESCRIPTIONS

(1) The Art of Philosophy

Philosophy is the art of explaining the obvious to create useful insights and perspectives.

(2) Degrees and Dangers of Evangelism

Faith requires commitment, and for any body religious to survive it must recruit, it must evangelize.

(3) A Dark Dilemma

Before dark stuff, our conceptual universe consisted of visible matter, four basic forces, and nearly empty space. Now, it consists of space filled with 72% dark energy, 24% dark matter, and only 4% visible matter.

(4) The Advent of Awareness

Myth and legend have clouded our vision since ancient times. Only recently have we begun to look past mythological and religious explanations for our awareness by shifting from belief-based perspectives to observationally based perspectives.

(5) Attractive or Compressive

Is gravity a compressed condition of space acting on matter, or an attractive force between matter acting on space, or both?

(6) When a Protest becomes a Threat

Confrontations over social issues can take many forms. Peaceful protests, obstructionism, intimidation, stalking, and violence, are examples.

(7) Tell the Fruth and Nothing but the Fruth

Fruth is what happens when fiction becomes entangled with the facts. Sometimes this occurs by accident, sometimes by being overly exuberant, but more often the twist is intentional.

(8) The Manipulators

Skillful detractors and disruptors can manipulate economic systems, legal systems, and governments.

(9) Translating to Reality

Language is a strange and wonderful thing. It connects us to others, makes civilization possible and, like the air we breathe, goes unnoticed.

(10) Are We Being Manipulated When We Vote?

The ability to measure, predict, and manipulate outcomes has become a science used extensively by political parties, lobbyists and campaign managers.

(11) Defending the Faith

When groups define who they are they also define who they are not.

(12) Weighing in on the Fermi- Hart Paradox

Planets and moons suitable for life orbiting other stars are too common for us to assume that we are alone in the Universe.

(13) The little Hiss

My interpretation of the creative event may not be a fully developed theory but should prompt a new perspective.

(14) When the Stars Speak

We are special but we are not free to ignore what the stars are telling us. They have waited a very long time for us to listen.

(15) Are Black Holes Hollow?

The paradox of black holes evaporating from their outer event horizons may be resolved if black holes also have inner event horizons.

(16) Digital Cloud or Digital Fog

The Digital age has transformed the way we communicate, the way we educate, the way we design, the way we build, how we manufacture, and the way we think.

(17) Infinities Large and Small

Infinities are enigmas that hide in the formulae of theoretical physicists waiting to pop up near the end of lengthy calculations making their efforts meaningless.

(18) Gravity <=> Acceleration

Treating time as a mental construct, added for convenience rather than as a quantitative constant, and by treating the speed of light as a universal variable dependent upon the expanded state of the universe, may resolve some recently discovered gravitational anomalies.

(19) Persistent Perceptions

The practical world of our existence is the result of our perceptive ability to interpret the rapid flicker of Planck moments.

(20) Vindication

The great minds of our time never claimed omnificence but I assumed it. Now that I know it isn't true, I can continue my own thought experiments and continue to blog on controversial matters with less intimidation.

(21) A Privileged Access to Reality

Our slow escape from assumed privileged insights and revelations has led us to an organized examination of the reality around us.

(22) Finding Order in a Push Pull Universe

Two men influenced our modern perspectives more than all others when they described the motions around us.

(23) A Very Dangerous Question

When gods were numerous and had many names one could ask about one's beliefs as simply and safely as if inquiring into ones health. Now, one risks a defensive response.

(24) A Civilization Built on Sand

The familiar metaphor that one should place the foundation of a house upon a rock rather than sand applies equally to a technology based on silicon.

(25) Militant Evangelism

Those who hope to save us from ourselves may be the most dangerous among us.

(26) Shaping Societies

Societies are defined by their history, their geography, their economics, and by the commitment of their citizens to the ideas that distinguish them as a separate and distinct group.

(27) Some Amazing Things about Language

"Admit it," he repeated. "The word of God is amazing." Without thinking, I responded in an equally loud voice. "What's amazing is that we can exchange complex ideas by vibrating the air between us."

(28) The (*Me–We*) Equation

The "Me/We" equation is weighted and balanced by circumstance but is perpetuated and observed as a matter of perception. It is sustained by the acquiescence of individuals to common perceptions.

(29) Inertia

The potential energy of motion, (inertia) and the potential energy of separation, (gravity) are more than cousins.

(30) Between the Past and the Future

The Universe may exist only as a succession of Planck units, a strobe lighted reality with a flicker rate far too fast for our slow pace of awareness.

(31) Describing Gravity

From the time pre-humans were being bruised by falling out of trees, we have been acutely aware of the force that tugs on us from below.

(32) Light Speed Illusions

Dark energy is not a topic of conversation appropriate in a grocery store check out line, but its cute title makes it interesting.

(33) Philosophy Ain't Dead Yet

Philosophy gets a bad rap nowadays because it is misunderstood, not because it is useless and boring,

(34) Bond of Awareness

We are more closely related to all living things by our common ability to sense and respond than we are by our genetic history of form and function.

(35) How Nature Builds Things

Humans create combinational directives that lead from raw ingredients to a product, Its what humans do, they build things. What about Nature?

(36) Legislate Don't Manipulate, A Plea to Politicians

Camouflaged campaigns recruiting and financing politicians in order to infiltrate legislative bodies for narrow personal agendas, is a subversion of democracy, not an exercise in democracy.

(37) Motion <=> Time

If everything stops, does time stop? Is time a requirement for motion, or motion a requirement for time, or both? Is time only a mental construct?

(38) Who Stirred The Pot?

As I get older, I occasionally feel a bit dizzy, especially after a martini. I attribute these periods of dizziness to having revolved with the earth 28,155 times since I was born.

(39) What's happening?

When someone asks, "What's happening?" and you answer, "Nothing much," you make a universal understatement.

(40) Size Matters

Does size have meaning only as a comparative that is relevant to the human scale of awareness?

(41) When Time is it?

We ask, "What time is it?" instead of, "When time is it?" as if we are asking for an explanation rather than to have a moment in time specified, Why?

(42) Killing Time

The movements around us keep our awareness busy and delude us to into thinking we are passing through time when we are only observing and recording synaptic intervals.

(43) Beyond Identity Theft, Identity Death

Using a laptop as a weapon, skilled information manipulators can have you declared dead and declare themselves your beneficiary.

(44) Why It Takes Two

Sex and death are intimately related. When life learned to share genetic material and enjoy it, it opted out of eternal life.

(45) An Army of Addicts

Armies enticed to fight by creating and controlling drug addictions is a disturbing concept.

(46) Creation's Patterns

We overlook the analog flow of creation because our conceptual capabilities have programmed us to look at and understand our surroundings piece by piece and one event at a time.

(47) A Question of Perception

Disagreements arise often over scarce resources and boundaries, but arise more often because of a difference in perspectives, personal, political or religious.

(48) Lines and Circles – Science and Religion

Many gods have dictated many standards for human conduct. Nature dictates her own mandates.

(49) Can One Dispute What Can't Be Proven?

Religious doctrines change as social organizations evolve and as doctrinal leaders reinterpret and add or remove passages from holy writ. Scientific doctrines change as theories are challenged and modified or replaced by new discoveries.

(50) PPD Politics

PPD politics have a long and dangerous history, have led to many atrocities and many wars and is now spreading across the globe under the guise of a Middle Eastern religion.

(51) A Message for the Press and Politicians

We like to think that our elected leaders and our free press are there to find expose and correct, or at least suggest corrections to, problems we encounter as a free society, but do they?

(52) Now and Then

For us, inside the universe asking questions, there is no *Now-and-Then*. There is only *Now* and an *Assumption*.

(53) The Ages of Language

My preference for historical categorization is to divide human history into periods separated by the different uses of language

(54) Alternate Explanations for Dark Stuff

Einstein did not like the idea of an unstable universe and inserted the extra term *Lambda* into his general relativity equation to keep the universe static. Later he called this his biggest mistake. I doubt if he would like the idea of dark matter and dark energy any better, but their introduction has put Lambda back into his equation.

(55) Can One Think and Emote at the Same Time?

In my youth I was instructed by a teacher, "You cannot think and emote at the same time". Until recently, I gave that instruction little thought, but prompted by recent terrorist acts and inflammatory political speeches I now find it relevant. There was wisdom in my teacher's admonition.

(56) Rare and Responsible

The power of awareness to affect its own future without reliance on genetic directives creates responsibilities.

(57) Turning the Universe Inside Out

Most of the universe around us escapes our attention. We assume things are solid because we can stand on them and they look solid, but when we investigate using scientific tools, things get strange.

ESSAYS

1

THE ART OF PHILOSOPHY

I entered college naive and uninformed. Other than the Bible, which no one in my family read, there were very few books in my home. I knew little about history, was the world's worst speller, had read almost nothing of significance and had no idea what I was choosing when I selected an introductory course in philosophy as a last resort to meet credit requirements. I expected the course to be a boring account of ancient Greek and Roman blogs and, in part, I was right, but there was more to the course than I expected. The dialogues of Plato and Socrates and the writings of Aquinas, Spinoza, Locke, Hegel and others contained interesting concepts that began popping up in other courses unrelated to philosophy. I learned that philosophy was orphaned when science became more inductive than deductive and, in the process of becoming more observant and less introspective, split into many disciplines. I continued to take philosophy courses and continued to have my perspectives adjusted as I searched for an all-encompassing definition for the collage of thoughts collected under the academic discipline called Philosophy. I never found one.

My career choices after college were more adventurous than academic but philosophical concepts stuck with me and resurfaced often. I now see philosophy as more of an art than an academic discipline and have my own definition.

I define philosophy as;

"The Art of creating useful insights and perspectives"

How we look at something is just as important as what we look at. This is true even in something as rigid as mathematics. New branches of mathematics, from number theory to tensors, are the result of new or adjusted perspectives. The same holds true in most areas of human knowledge. The window we use to view our surroundings frames and limits our impressions and we have many windows from which to choose. I see the philosopher as a guide, taking people from window to window, pointing out obvious differences in various views. The philosopher is also an artist that uses words to paint perspectives by arranging synaptic connections, and by raising questions that linger, questions the philosopher prompts, but rarely answers.

Hemlock anyone?

2

DEGREES AND DANGERS OF EVANGELISM

In the Bible, faith is described as, "a commitment to things unseen". Before one can commit, however, the things to which one commits have to be explained and learned. If one doesn't know of Buddha's teachings, they cannot follow Buddha's advice. If the writings of Joseph Smith are not explained, the Mormon faith cannot be followed. Similarly, the teachings of Christ or Mohammad cannot be followed until one has been exposed to and indoctrinated in the tenants of the faith. Faith requires education and commitment, and for any *body religious* to survive it must recruit, it must *evangelize*. Even monasteries reach out to recruit and train future monks to insure the future of their spiritual sanctuaries.

Evangelic messages all have common elements in spite of divergent beliefs. They all proclaim their faith to be the best or the only true path to a relationship with a god and immortality. Hidden in any evangelic message is the statement; *"You are missing the truth, need to be corrected or informed, and it is my duty, according to my faith, to challenge your beliefs, or lack thereof, and save you from your delusion."*

This need for the religious to challenge other belief systems is innate in all evangelical activities and creates deep lines of distrust between groups of people trying to live together in assemblies created by other necessities, geography, migration, and economics, for example. In this sense, the evangelical, no matter how sincere and peaceful their intent, are creating dissent, and the level of dissent is directly proportional to the emotional level of their evangelic activity.

The sacred directives of the world's dominant religions all contain directives for confrontation that clearly distinguishing between true believers and infidels, (the faithful and all others). These directives are in the Torah as Abraham is directed to destroy all others in the Sinai to make room for the twelve tribes of Israel. They are In the teachings of Christ as he excludes the uncircumcised as unworthy of Gods grace, In the papal bull of Pope Nicholas V declaring slavery an appropriate state for non believers, and in the Koran as it declares death to the enemies of Islam being a responsibility of the committed. Human history is filled with wars and genocides generated by evangelical fervor and only recently have Man's evangelical confrontations been partially set aside because a few wise men followed the advice of John Lock in his 1688 "Letter Concerning Toleration".

John Adams, Samuel Adams, James Madison, Thomas Jefferson, Patrick Henry, Benjamin Franklin, and others, after much debate, brought the colonies together by putting aside their many religious differences and established the first secular national state. Under this arrangement, matters of state are settled by reasoned debate without the intervention of religious tenants. The state, in turn, is restrained from interfering in church matters and is pledged to protect the right of all citizens to choose their religion without coercion. Most European countries have now adopted this separation of church and state as beneficial to both the state and religious freedoms and, until recently, wars have been fought over other ideological differences. Unfortunately, evangelism has once again risen to a dangerous level and coupled with mass media and the internet is inspiring radical acts undermining the bulwarks against theological intrusions into the separation of church and state.

Faced with a militant form of evangelism in the Middle East, the faithful in the US have responded politically by inserting protestant Christian beliefs directly into government through gerrymandering and well-financed primary elections. These covert evangelical political campaigns are as dangerous and divisive as any other evangelical effort and have the potential to undermine our constitutional system. If they succeed, the religious will undermine the secular state and loose the protection the secular state provides them. If they gain control of government, their dominant position will be challenged by both the secular and other belief systems. The goal of Isis is to create a religious war. The most foolish thing we as free people can do is to engage them by undermining our secular government traditions.

The irony of young Mormon missionaries from the United States, being injured, in Brussels by radical Muslim missionaries when they bombed the airport exposes an extreme difference in evangelical methods but identifies a similar evangelical goal; "go abroad and recruit".

Evangelical efforts can be,

- *passive,* as in leaving the windows of a church open to allow passers by to hear the sermon,
- *moderately active* as having a quiet religious discussion at Starbucks,
- *active* as in passing out literature, knocking on doors, or sending missionaries to other countries or,
- *militant* using intimidation, imprisonment and executions.

Any form of evangelism, religious or other, establishes a *We / You* distinction that is divisive and dangerous, especially when individuals or other groups of believers are told their beliefs are false and need modification. Evangelism is a social irritant of significance and the *good* the evangelist assumes they are doing may actually be a major cause of violence and suffering. As a way of governing, rational discussion and compromise are the alternatives the secular state offers to governing by tenants of faith beyond argument.

Isis will be with us as long as hypnotic religious diatribes continue to turn rational individuals into emotional robots. Until we understand how religion is able to override reason, we will not understand the radicalization process, and won't be able to defend against it. Religion and reason have been at odds since serious investigations into natural processes began. Investigations into the effects of religion on emotional responses and survival instincts have been off limits as a breach of god's spiritual connection to humans, but as we examine nature, god's benevolence appears to have been subverted or misunderstood. It is time for the leaders of the faithful and the leaders of citizens to begin a serious dialogue to discuss, *rationally*, the relationship between religion, science and secular governments, and to acknowledge the common dangers both face from subverted evangelism's ability to destroy what the cooperative efforts of humanity have built.

3

A Dark Dilemma

Scientific theories are descriptions. They are not explanations. Scientific theories develop slowly and only a few create new and useful perspectives. Evolution and Relativity are examples of theories that have refocused our perceptions.

Successful theories affect more areas of human endeavors than just the direction of scientific research. They alter perspectives within the general population. Successful theories sometimes give us a sense that we have arrived at final answers, but they can also make us question even the most established scientific paradigms. Measurements of galactic movements, starting in the 1930's created such questions and have prompted new theories, especially regarding gravity.

Gravity has been elusive in our efforts to match it with the other three basic forces of nature. The recent discovery of a 'Higgs' particle renewed our hope that we might be close to a unified theory explaining the connection between atomic actions and galactic movements but, instead, the discovery of this large new atomic particle created only new questions.

Newton described the movements of stones falling to the ground, artillery shells curving in flight, and planets orbiting the sun as the result of an attractive force between particles of matter. Einstein described the same movements as the result of matter curving space. Both men used innovative mathematics. Newton used the calculus. Einstein used tensors. Both men assumed that matter influences motion, Newton by having matter act on itself, Einstein by having matter alter the geometry of space. Einstein's relativity replaced Newton's mechanics by using a curved space theory to reconcile the planet mercury's observed orbital discrepancy and by predicting altered paths of light passing near massive objects. Observations dictated the final choice between theories but neither theory accurately describes newly discovered movements at galactic scales.

With only one macro attractive force in our conceptual repertoire we continue to use it to explain the anomalous rotation we observe in galaxies, even when there is no visible matter to explain it. Unwilling to question Newton or Einstein, we calculate how much missing matter is needed to cause the strange rotations we observe and conclude it must be hiding.

The result of our inverse analytical calculation is a massive amount of invisible matter embedded in and around a galaxy's visible matter. Choosing to explain the anomalous rotation of galaxies by additional gravity, and with no way to explain the gravity without additional matter, we have concluded that an invisible form of matter has to be the cause when all we really know is that galaxies don't rotate as we expected. As a result, we have focused our research on finding and identifying missing dark matter. However, there may be other ways to explain the observed errant galactic movements.

In 1983, Mordehai Milgrom formalized a modification of Newtonian dynamics that describes the anomalies observed in galactic rotations as natural variations in gravitational and accelerative forces acting on galactic and larger associations of matter. His predictive formulae also accurately describe the movement of double stars, satellite galaxies, interacting galaxies, and accelerative forces. Unfortunately, scientific research has a momentum coupled to long-term research grants and academic inertias that can sidetrack promising theories.

The anomalous rotation of galaxies was first noted in the 1930s and has been studied in more detail ever since. The idea of hidden matter producing additional gravity is not new but the name *dark matter* has appeal and attracts media coverage and research funding. The *Modified Newtonian Dynamics Theory, MOND*, applies to the same observed galactic phenomena but without any dark postulates and, unfortunately, without the same media appeal.

Our conceptual universe was once filled with visible matter, four basic forces and nearly empty space. The dark world of science is now filled with 72% dark energy, 24% dark matter, and only 4% visible matter implying that after more than two thousand years of exploration, we understand less than 4% of what is going on around us.

Searching in the dark for matter and forces hidden from us, we are reaching beyond current conceptual limits and beyond our most basic assumptions. *MOND*, the modified Newtonian theory, offers us an alternative that keeps us in the light without the need for dark matter and brings us closer to reconciling gravity with the other forces of nature.

Our primary sense is vision. Visible light has always been our most important portal into the workings of nature. We followed the motion of celestial objects with our eyes noting their movements until we could track and predict their paths. We observed the similarities and differences in living forms, detailing their shapes, colors and behavior in books and paintings until we could explain their similarities as the result of a common trace. We have also developed tools to convert sound and other vibrations into visible curves for analysis. We have invented cloud chambers to study the make up of matter using visible traces of particles too small to see, and have extended our understanding of light to frequencies beyond those available to the eye and now use microwave and infrared wavelengths to our advantage. We also use symbolic languages, verbal and mathematical, to create descriptions of our observations and predict future observational opportunities. Our dependent relationship to light and the pervasiveness of the electromagnetic forces throughout the universe has formed our basic perspectives, shaped our questions, and guided our investigative efforts.

Concentrating on *(invisible matter) and (invisible forces)*, instead of re-examining gravity leaves our scientists nearly impotent. Dark matter and dark energy theories find little footing and few experimental options. Alternate theories that allow us to continue to searching in the light may also have implications for relating gravity to the strong force and creating a unified theory.

We seem to be searching in the dark for answers that are waiting in the light. My money is on *MOND* and similar theories with a trillion neutrinos tossed in for balance.

4

THE ADVENT OF AWARENESS

Myth and legend have clouded our vision since ancient times. Only recently have we begun to look past mythological and religious explanations for our advanced awareness by shifting from belief-based perspectives to observationally based perspectives. Historical events, long misrepresented by contrived narratives, are now being described more accurately using investigative methods. Human history is no longer measured in *biblical begets*. It now stretches back past mythical gardens to connections with the earliest forms of life and our advanced intelligence back to the first faint flickers of awareness.

Time is no longer limited by the human pace of awareness. Man is no longer the center of the universe and our standing as an intelligent creature no longer requires a personified god to explain. The story of man is no longer one of mythical miracles but one of natural processes. Human history is no longer a short and simple story. Instead, it is a long and complex series of adaptive occurrences. Man, as a species, is the most recent spark in a chain of living energy begun by chemical reactions four billion years ago and man's advanced awareness is the result of millions of trial and error adaptive events, not the result of eating forbidden fruit.

To visualize the time needed for the long adaptive process leading to mankind, picture yourself' sitting in stadium seat #1A of a large sports arena. Sit there for a full day and then move to the next seat for a day, and then the next and so on until you have sat for a day in all of the thirty thousand seats in the stadium. Then move to another stadium and continue until you have sat in every stadium seat in the world. The time involved to accomplish this task is so lengthy that sports events will change, the design of stadiums and stadium seats will change and you must not only adapt to new seats but to new stadiums and new sports events. To further challenge your efforts, new stadiums are being built faster than you can sit in them. You will never catch up and will never be finished.

The stadiums in this analogy are environments that are constantly changing. The time required is the billions of years needed for evolution to take place. Living things change slowly from generation to generation, "seat to seat", and must produce progeny that fit new environments, new seats and new stadiums if they are to continue. Only by chance genetic deviations creating offspring that fit new seats and new stadiums can the genetic trace continue. Offspring that remain unchanged are stuck, and disappear when the stadium is torn down.

Humanity's successful history in this analogy is one of parallel adaptations, two sets of progressing adaptations that have compounded the survival advantage of humankind. Genetic errors have played a major role in squeezing us out of the early living soup of single cells and allowed our ancestors to adapt to environmental changes including,

- continental drift,
- ice ages,
- volcanic catastrophes,
- and meteor impacts.

Changing environments have weeded out and sorted through millions of life forms since the oceans formed and, against all odds, the seed of primates made it through the evolutionary sieve and persisted long enough for a few to develop traits needed for their continued survival. Among these traits were the precursors for complex vocalization and attentive situational evaluation, *speech, cognitive awareness, and memory.*

Enabling and reinforcing these essential traits were fortuitous physical adaptations allowing upright walking and the ability to manipulate small objects. From these combined evolutionary opportunities humans emerged with communication skills, imagination, enhanced learning abilities, and the ability to manipulate the environment to insure their survival.

Our advanced awareness is the result of early sensitivities being carried forward in concert with the evolution of complex forms. When microscopic life became mobile, a sense of surroundings became essential, and as multi cellular forms moved into increasingly diverse environments, keen senses and learned responses became even more advantageous. Advances in awareness that kept pace with advancing physical abilities further improved survival rates.

Complex living forms and advanced levels of awareness are a favored survival partnership. Humans are a direct result of this partnership and the product of the long periods needed for matter to become animate and awaken.

Shortly after the earth cooled and the oceans formed, a stable atmosphere developed and the awakening of matter began. The lipids, salts, and amino acids necessary for life were abundant and life self-assembled from the mixing of common materials.

The first living cells led to a long history of cell divisions with each division forming identical sister cells. Early life was homogeneous. Only an occasional error in replication created new results and the environment measured these new microscopic attributes for survivability by favoring the best adapted. Altered replication machinery inside every mutant cell was also tested to determine if a new attribute could be duplicated. Chance mutations had to pass both tests before becoming a part of the ongoing chain of life. The process of change was slow until cellular complexity reached a critical level, (eukaryotic cells), and the process of errors and testing accelerated.

Changes in mitotic genetic replication depend on errors, (mutations) and as cells grew in complexity, duplication directives became more complex, and were more prone to errors. With more complexity came more errors more testing and more mutations. This self-accelerating process led to a progression of complex directives, more complexity and the first multi cellular organisms. The long process of single celled life emerging from natural elements included:

- The development of a cellular membrane and DNA (prokaryotic cells)
- Learning to collect and cooperate in mass, (mats and stromatolites)
- The development of internal organelles, (becoming eukaryotic)
- The development of basic sensitivities to heat, vibration, various chemicals and light, (the precursors to sensory organs)
- The ability to turn carbon dioxide and sunlight into chemical energy, (photosynthesis)
- Adapting to survive in an oxygen rich atmosphere after poisoning the environment with their own emissions (The first self induced ecological disaster)

- The ability to transfer combinations of genetic directives to others and the production of multi cellular and multi organ structures, (meiosis).

During the billions of year's single celled life was developing complexity it was also laying the groundwork for the expansion of awareness. Our instinctual ancestral awareness began in our single celled ancestors and continued through creatures with rudimentary backbones, amphibians, early mammals, the first primates, and early humans. The structure and function of our multi component multi layered brain reflects the contributions of these ancestors. Our reflexes, instincts, and emotions are, in large part, inherited traits with roots reaching back along a genetic trace starting with the first replicating strands of amino acids.

With the advent of meiosis and multi celled organisms genetic adaptive experiments increased exponentially producing millions of new life forms. Complex cellular arrangements were tried in great numbers but the ways in which any living form can sense and react to its surroundings is limited. Most experiments with varied forms have failed, but common sensory pathways for evaluating one's surroundings and one's internal conditions have survived and advanced. These limited pathways, (light, vibration, chemical, pressure, temperature and electro magnetic), are the foundation for all (sense and react) systems and are common to all living creatures alive today. How we perceive our surroundings and ourselves is innate, patterns established long before we became human.

Humankind, as a species, is a recent arrival in evolutionary history that brought with them a combination of neurological and physical attributes that has altered his home planet and taken control of the evolutionary process. Man's surge from a common primate to an animal capable of examining the structure of the universe and the makeup of matter is the result of an explosive advance in living awareness from animate and reactive to analytic and self-directive.

This leap from inanimate to aware, of stardust arranging into a complex investigative organ and becoming aware of itself, is a significant event in the evolution of matter and the universe.

The advent of human awareness is an event as significant as the advent of atoms, the formation of stars, and the super novas that created the dust that now analyzes itself.

As individuals, we are essential but insignificant. As a cooperating cloud of awareness, our significance is much greater than we have begun to appreciate. Beyond religious commandments and human dictates Nature and our advanced awareness imposes three greater mandates:

To explore and learn
To be of good council
To be a good steward

5

ATTRACTIVE OR COMPRESSIVE

I see the leaf twisting in the air but cannot see the wind and yet I know it must be there. When I stand, I feel the weight upon my feet and see the stone drop from my hand, but cannot see the force that makes it fall. Sight does not reveal everything. Many unseen forces direct the dance of matter that we measure. We name these forces and predict results as they direct the dance, but our descriptions never generate complete answers, instead, they generate more questions.

Our latest observations have generated two new questions. One of the questions we have answered by assuming the existence of invisible matter. The other question we have answered by assuming the existence of an unexplained new force. With no observable attributes to use as naming clues, we have named them *dark matter* and *dark energy*.

Both new entities have been exposed by the anomalous behavior of gravity. A known force we have described well, but never explained. One of the new dark entities appears to add gravity. The other seems to oppose it.

As theorists seek answers to these new questions, I continue to seek answers to older ones, specifically; is gravity attractive or compressive or both, and is there a difference? Before you tell me to put my hand down as if I were in a classroom asking a dumb question, let me explain why I am confused.

I have been told by experts that pressure inside a star or planet increases with depth below the surface and is greatest at the center. This seems correct if celestial spheres are being squeezed into shape from outside, but if they form by atoms attracting each other, it would seem that pressure would increase with depth until the attraction of the atoms above begin to balance the attraction of the atoms below.

I liken this to a tug-of-war between the attractions of atoms below competing with the attractions of the atoms above. On the surface of a massive sphere, all of the atoms are below with all of them attracting things on the surface more or less toward the center. Place something deep below the surface and some of the atoms are now above attracting it toward the surface. Place the item at the center and there are only atoms above, pulling the item toward the surface in all directions leaving it weightless. We know this isn't true. But why?

If gravity can be explained by the attraction of atoms being additive, a zero gravity and zero pressure condition should exist at the center of planets, stars and black holes, but this contradicts observation and I remain confused.

Is the nickel, iron core of the earth floating in a low gravity state contained by a compressive mantle, or does the gravity vector continue to the center of the earth causing the pressure to increase to a maximum at the center?

If we accept the premise that pressure and gravitational forces inside a massive sphere increase all the way to the center, the compressive condition would seem to be directed toward the center by something other than the additive attraction of atoms.

Einstein described gravity, using tensor graphics, a four dimensional representation of a formula equating mass with distorted space and paths of movement. The descriptive power of his formulation has usurped both Galileo's and Newton's descriptive formulas as they apply to massive objects but, If macro gravitational attractions are the result of atoms altering the geometry of space around them, how are these micro spatial distortions added to produce the macro effects observed? Distorted gas clouds in space appear to fall, or be pushed, toward their center of mass, eventually gathering as tightly as they can around a center point creating a shape with the smallest surface area possible, (a sphere).

Space distorted by matter was Einstein's way of describing gravitational effects. What Einstein left unanswered, was not the macro interrelationship of matter and space/time but *how* matter alters space. Is gravity the result of space compressing matter within it, or are atoms causing the condensed condition?

Newton described gravity as an attractive force between bits of matter and even devised a formula for a zero gravity condition existing in the interior of massive hollow spheres. Einstein's field theories describe a connection between matter and gravity similar to a couple arguing about who is in charge. Mass is pulled by Gravity but when Mass complains, Gravity responds that Mass is initiating the pull. A resolution of who is in charge of this relationship, (mass or gravity) seems impossible, and divorce doesn't appear to be an option.

Einstein's general theory has been accepted as the better description. Sub atomic particles with mass may curve the space around them, but is the curving effect on space cumulative as mass gathers always creating a center of mass and a point of greatest curvature, or do minute gravitational space warps interact in a Newtonian way. Observation and experiment point to the former but a better description of this cumulative effect would help resolve my confusion.

One approach might be to consider space as having an energy density different from the density of the matter within it. An atom is mostly empty space and like the empty space between atoms has an energy density. Atoms in space create a space within a space and create an interaction between these spaces dependent upon their energy density difference. Much like a helium filled balloon seeking a density balance by ascending in the atmosphere, or a beech ball rising rapidly to the surface when submerged in water, atoms with an energy density less than the energy density of the space around them seek a density balance by moving toward an area of lower density.

If, as Einstein suggests, matter affects the space surrounding it, but instead of curving it, reduces its energy density, the results are the same, but the explanation is different. If atoms and congregates of atoms have a *low energy density* as compared to a *higher energy density* for surrounding space, there exists a buoyant potential for atomic matter to move away from high density areas of space and toward the lower density areas they create.

If matter alters the density of space instead of curving it, Einstein's *feldgleichengen der gravitation* is still valid but makes more sense and the compressive effect of gravity becomes an interaction where a formed atom with a low energy density reduces the high energy density of the space around it by absorbing energy. Gravity then becomes a buoyant force with low-density atoms migrating toward a low energy density state in the space near them creating a low-density center we call a center of mass.

. This micro / macro perspective of an inverse-buoyancy in space may also shed light on our new dark dilemmas and fits into Einstein's equation describing gravity with the simple substitution of a negative density constant for his gravitational constant. It may also be described by combining simpler equations for density, volume, and buoyancy to describe the tendency of matter to join and compress in space.

Visualizing the inverse buoyant concept for gravity involves thinking of space as empty of matter but full of energy, of having a zero material pressure but a positive energy pressure, (an energy density). It also involves thinking of matter as visible energy made solid and interactive by electromagnetism. We are close to these conclusions in our dark energy explanation for an accelerating expansion of the universe and in our description of the massive components of hadrons, as composed of 99% energy density with a residual strong force field around them.

The motions and pressures we ascribe to gravity can be described by an inverse energy dense buoyancy If,

1. Point energy particles called quarks and the energy state that joins them into a mass particle is where gravity begins.
2. The joined quark energy state has an energy density less than the average energy density of space.
3. Differences in energy density produce buoyant effects that mirror buoyant effects in macro density situations by seeking an equal density position in a density gradient

Think of a single proton as a fuzzy energy bubble in an energy field called space. As these low density bubbles attract each other toward their low-pressure centers they join creating a larger bubble that reaches out to draw in more atomic bubbles, all of them pushing toward a low-density center by buoyant forces. The solids with mass we observe attracting each other and joining are, in fact, low energy-dense states formed from the residual effect of joined quarks.

Matter looks and feels solid but is really only electromagnetic interactions created by point energy particle waves we call quarks and electrons. Our reality is a synaptic reality created by organic adaptations to a world formed from the interaction of four basic forces held together by an energy density difference. We can't see it but we can see its effects and can deduce its nature. How we look at things is just as important as what we look at, and our scientific theories are only a current perspective that best matches our current observations. Finding a pervasive universal energy may lead us to adjust our view of gravity and come closer to simplifying and joining our micro and macro theories.

If gravity is the buoyant effect of an energy condensate called quarks and their joined energy state called hadrons floating in the energy field we call space, the pressures that fuse atoms in the center of stars and form black holes are the result of buoyant pressures created by a difference in energy densities.

6

WHEN PROTESTS BECOME A THREAT

Confrontations over social issues take many forms. Peaceful protests, obstruction, intimidation, stalking and violence, are examples. Unfortunately, laws do not adequately differentiate between these various confrontational forms. In free societies, laws do not always treat ideological differences fairly for fear of trampling on a citizens right to free thought and expression. In more despotic forms of government protests are dealt with swiftly and sometimes with deadly force. Free speech is an essential element of democracy and guaranteed by our constitution, but there are limits. Slander is not allowed. Creating panic without due cause is not allowed and certain activities used as a means to voice an objection or opinion are not allowed.

One cannot jamb up a revolving door to a department store because they sell perfume tested on animals, or drop stink bombs onto a construction site because forests are being destroyed. Striking workers however, can recruit surrogates to parade up and down sidewalks with ready made signs, war protestors can infiltrate funerals for fallen soldiers, religious protestors can interfere in gay right activities, protestors can block access to abortion clinics, and neo Nazis can parade with anti-Semitic banners.... *as long as these activities are infrequent and of short duration*;

There is, however, one ongoing protest that has lasted for decades and is an every day occurrence. It is carried out by the same few individuals who claim their right to protest cannot be restricted because their message is from God. They claim they are free to continually harass, intimidate, threaten and promote harm to others with impunity. Clergy, politicians and lawmakers cower and avoid suppressing these protest activities because this small group has succeeded in drawing the line between what they believe, and any modified or alternate view so narrowly, that any reasonable attempt to restrict their activities identifies the police or judges as non-Christian. These same militant few are treated with kid gloves by police and prosecutors because the protestors revel in anyone's attempt to control their activities as an opportunity to publically broadcast their obstructionist cause, and claim they are victims of state interference in the free expression of religion. Any other continuing protest activity lasting decades and involving stalking, personal threats, invasions of privacy, bombings and killings by a small group of militant evangelists would have labeled them terrorists and would have been dealt with swiftly. However, these few radicals have intimidated our leaders and infiltrated government agencies, not to serve the people but to serve a single obstructionist cause. They succeed because their voices are loud and threatening, not because they truly understand or are compassionate. They persist because they have used their single issue to divide our society into believers and non-believers and have created a state of avoidance by the law. Ignore them and they will never go away. They have even corrupted our language.

Before their decades long campaign of intimidation, the

word *choice* had a good connotation. It was at the core of our democracy and morality. Now the word has been turned into a single evil option that cannot be used without their agenda ringing in our ears. I would urge everyone even those partially seduced by their rhetoric, to go unannounced to one of their protests and pretend you are violating their barrier, or just stand aside and observe. These people are infiltrating your government and slowly usurping your rights, not only as it pertains to their cause, but many more.

The intensity of one's beliefs does not make their beliefs any more or less true, and this type of militant evangelism is potentially as dangerous as any other radical cause. I would ask those that enforce the law to apply it equally to these obstructionists. A thirty-year reign of intimidation and stalking is not a peaceful protest. I would urge the rational clergy to disclaim these religious intimidators and allow others with different beliefs to exist in peace. If you ignore them, you condone their approach and will eventually draw their militancy into your churches and your living rooms. We have had enough of religious wars and the deeper you draw the line between those who pray and those who don't, the less likely anyone seeking religious refuge will cross the line, and enter a church.

7

TELL THE FRUTH AND NOTHING BUT THE FRUTH

Fruth is what happens when fiction becomes tangled with the truth. Sometimes this occurs by accident or by being overly exuberant, but more often, this twist is intentional, and when used by skilled language manipulators, it is difficult to identify. As a carefully concocted almost truth, fruth can misinform, mislead, and malign. It can sway elections, warp important decisions, turn a benign religion into an army, create mass hysteria and destroy nations. Fruth has caused witches to be burned, the innocent to be executed and allowed sociopaths to acquire high office.

Politics is filled with fruth as differing opinions and opposing ideologies clash in the contest for public opinion. In 2016, abortion again became a flash point in the US congress when the religious right tried again to blur the line between church and state by pitting the bible against the constitution. Facts in such arguments are not clear and concepts are easily fruthed. In this case party "R" threatened to hold the national budget hostage if an anti abortion provision was not included. This forced party "D", attempting to keep the budget process clear of such side issues, to reject the infected bill. Then party

"R" who originally threatened to shut down the government accused Party "D" of being the cause of the near shut down by not acquiescing to party "R's" demands.

It doesn't get much fruthier than this ridiculous display of twisted facts.

8

THE MANIPULATORS

Of the three primary mandates imposed on us by our advanced state of evolutionary awareness, *To Be of Good Council,* may be the most important. Our advanced language capabilities define us as a species. We alone can formulate and exchange complex ideas verbally and in writing and the results of these exchanges create our social arrangements, our governments and our laws.

Language, even in its many forms, will never be able to capture all of reality. Our mathematics fails when descriptive limits become infinities. Our speech fails when words prove insufficient to carry an idea or when they are manipulated to mislead or misdirect. Words define and structure our most important organizations and are sustained by individuals who understand the limitations of language and the fragile nature of the organizations created by men and women of good council. The institutions that create order in our lives are built on processes described by our words. They only work well if the processes described are followed and allowed to evolve through considered debate.

Being of good council requires knowledge, understanding, forethought and effort. To think before one speaks and to listen with an open mind while being mindful of the forum in which the exchange is taking place, are the courtesies that allow civilizations to form and continue. Unfortunately, economic systems, government processes, and legal systems can be manipulated by skillful detractors and disruptors. Legislative manipulators use committee positions to sideline important debates and flood the debate floor with useless bills and speeches to avoid regular order. Sophisticated investors manipulate the stock market to make millions through short selling, lawyers manipulate jury selections to insure verdicts, banks manipulate lending rules to take advantage of a gullible public, wealthy investors manipulate elections to insure outcomes, corporations manipulate suppliers and shippers to insure greater profits, and the wealthy manipulate tax laws to avoid taxes.

Manipulation has become so rampant in our capitalistic democracy that it is flagrantly practiced in our institutions, in business and in the stock market. Manipulation has become ingrained and accepted, but it erodes the institutions it feeds on, weakens the economy and the nation. Reason and restraint sometimes intervene but are but whispers in the shouting contest for advantage. Have we become a nation of competitors, forgetting that it was cooperation that made us great. Have we become a nation of profiteers instead of patriots, a nation of manipulators instead of managers, or are we just undergoing a temporary setback in reason and civility?

9

TRANSLATING TO REALITY

Language is a strange and wonderful thing. It connects us to others, makes civilization possible and, like the air we breathe, goes almost unnoticed. For humans its primary essential carrier is sound, (vibrations in the air). Sensitivity to vibrations became a valuable survival attribute very early in life's slow evolution. Even before living cells began to join and share functions, early bacteria developed primitive methods of monitoring and responding to vibrations in their immediate surroundings.

Evolution is driven by small advantages introduced by chance and are carried forward in nature's evolving genetic language. As nature's language evolves, it instructs each host cell how to grow and how to interact with its neighbors. Instructions that degrade the host cells ability to survive also reduce the chances for survival of the genetic pattern giving the instructions. The language of life is modified as its instructions produce more, or less, sustainable living forms.

Genetic patterns directing improvements in detecting subtle differences in vibrating patterns have proven to be valuable and have been continually replicated and improved. After three and one half billion years of testing, the test results for sensitivity to vibrations are in. The results are complex ears and advanced synaptic connections for interpreting sounds. However, sensitive ears did not evolve alone.

A parallel evolution of mutually beneficial traits is common, especially in multi-celled organisms as they struggle to adapt to rapidly changing conditions. In humans, the parallel evolution of an attuned ear with an articulate tongue proved especially beneficial. Being able to interpret sounds and replicate them and to use vocalized sounds to describe things or to give directions, gave humans an enhanced ability to cooperate at a higher level and an advantage over competing species. At first, simple human languages were composed of only a few descriptive nouns and a few calls to action, but they quickly developed into a more powerful tool.

Language advantage gave man an edge in survival wars that promoted further refinement in the genetic instructions for more sensitive ears and more articulate tongues. At first language was simple and only spoken, but with the first rough depiction of an animal scratched on a flat stone, language took on another dimension. The verbal sound coded to match a real animal was now also associated with a symbol and the symbol could evoke the associated sound. With this fortuitous step, reading and writing were born.

At first, the scratches were primarily of objects with very few showing action, but as more objects were depicted common symbols developed becoming precursors to both nouns and verbs. Tribal members capable of looking at the symbols and translating the marks into a verbal account were held in high regard. They made the scratches on stones talk and to early man the process seemed magical.

Human language has since evolved to accommodate more than objects and actions. It now encompasses ideas shapes, numerical relationships and fantasies. The scratches have also become symbols for sounds that depict ideas that have little or no pictorial content. As the literate class grew, many learned the trick of making the written symbols talk inside their heads, and the ability to read was no longer held solely by priests and the privileged.

Learning to read is now an essential ability necessary for society to function, but the translation process from symbol to sound hasn't changed. We learn whatever language we speak as a child by listening to those around us. In school, we learn to represent vowel and consonant sounds on paper by making scratches with a pencil. Then we learn to translate the scratches back into sounds and make the scratches talk. Eventually we learn to let the symbols talk inside our heads without verbalization, eliminating the need to vibrate the air with tongue lips and vocal chords in order to communicate.

The silent method of communicating with scratches has advantages and disadvantages. Its disadvantage is that it is not immediate and is inappropriate where spontaneity or a quick response is required. Sending a note to someone instructing them to "Get Down" when a bomb is about to go off, or texting "LOL" when someone tells you a joke at dinner are obvious illustrations.

Vibrating the air between us is still an essential method of exchanging information in casual conversations, when making public speeches, when cheering at a sports event, or when calling your grandmother on the phone to wish her a happy birthday.

Scratch communications, (writing), has the advantage of permanence. On the book shelf in front of me is a copy of "The Iliad". The book is quietly waiting for me to open it and let Homer speak from the past without vibrating any air. As long as I can translate the scratches on the page, my inner voice will speak for Homer and for hundreds of thousands of other writers, all waiting to speak with the simple turn of a page. Learning to translate symbols to sound, blueprints to a building, instructions into actions, and ideas into reality, is the trick that sustains us and makes us superior to other life forms. It also makes us the most dangerous. Language allows us to mislead ourselves, and others, by distorting reality, by creating make believe worlds, by inventing Gods, and by using twisted logic to convince ourselves that what we hear in our heads needs no verification.

Of the three mandates evolution has placed on us, (to explore and learn), (to be of good council), and (to be a good steward), being of good council, respecting the gift of language, may be the most important as we translate our beliefs, our ideas, and our attitudes into reality.

10

FOR WHAT ARE WE REALLY VOTING?

Science and math now play major roles in the politics of democratic societies. The ability to measure, predict, and manipulate outcomes has become a science used extensively by political parties, super-pacs, lobbyists and campaign managers. The actual beliefs, character and issue positions of candidates have become secondary to carefully crafted campaigns, prepared statements, and canned responses. Candidates for political office have become created personas molded to get desired responses from carefully measured attitudes of the voting public. The individuals, shaking hands and making speeches have become disguised, coached, and prepared images more than real persons.

Candidates are chosen for name recognition, public status, and their measured appeal to specific portions of the electorate. Carefully vetted candidates are more like actors on stage in front of elaborate backdrops than true individuals seeking a leadership role and asking for voter trust and approval. Election poles and results have begun to give all voters pause and make them question, not whether the election was rigged by manipulating vote counts, but how much, each of us may have been manipulated by statistical analysis and slick advertising to make directed selections in the voting booth.

The same science that manipulates our politics is also used to manipulate our shopping choices, our social behavior and to sway trial outcomes by a careful selection of jurists. Innate responsive behavior by individuals and groups has always been with us, but until recently, it was unmeasured. Until recently, the outcomes of group decisions seemed random and somehow fair. Political candidates were real people asking for our vote, our choices of products a thoughtful selection made of our own free will, Juries, composed of individuals with random opinions were thought to meet out fair and impartial justice, and election results were thought to be the product of group wisdom.

These assumptions may have been naïve but we felt empowered by our freedom to choose and took pride in the expression of our individuality. Now, we leave the grocery store or the polling booth wondering if our choices were really our own. Have we becoming sheep, herded subliminally by teams of big brother corporations and rich individuals? The uncertainty raised by the power of statistical analysis echoes all around us, and our distrust is making us even more susceptible by the paranoia it creates.

Some of those playing on paranoia are simply unethical radio hosts using public uncertainty to improve their ratings and income. Others are pushing our emotional buttons to gain control and change the way our democracy works. Freedom is not simply freedom from big government, it is also freedom from the imposition of religion, freedom from manipulative political indoctrination, and the freedom to participate openly in a diverse cultural mix under the umbrella of a compassionate government that mollifies extremes and recognizes everyone's contributions. The only way to keep the sheepherders and their statistic dogs at bay is to think for ourselves, to begin to question our programmed emotional responses and to separate ourselves from long and persistent indoctrinations by family, friends, talk radio, church, and slick advertisements.

Induced political paranoia serves no purpose in a democracy. To preserve our form of government we need to use reason instead of preprogrammed emotional responses and must expose and avoid extreme views and positions. If we are to take back democracy and keep our freedom, we must learn to vote from an informed position, not as angry puppets.

11

DEFENDING THE FAITH

Be it a faith based organization, an ethnic group, a Nation State, or a common set of cultural beliefs, groups define themselves, not only by stating who they are, but by emphatically stating who they are not.

Any organization that is threatened by outside influences or is unwilling to compromise its basic beliefs or principles, continually expends energy and assets to maintain its integrity. In these organizations, alternate systems and beliefs are viewed with caution or as a threat. Self-defining systems can create cultural tensions and conflicts and religions are not exempt. History is full of conflicts and atrocities caused by diverging beliefs within a system, and as the world becomes more secular, religions become more defensive.

With few offensive options available, beyond converting non-believers, some religions go to extremes and attempt to eliminate or neutralize the threat. A devout individual or organization unable to accommodate alternate perspectives can easily become frustrated when their belief system is questioned threatened or compromised, even when the threat is imagined. This impulse to take defensive action can also be intentionally triggered by powerful or charismatic leaders and has been used often to create chaos, suicides, and killings on massive scales. When a set of beliefs become the primary definition of self for an individual or a group of individuals, paranoia is common and violent reactions become possible.

The Middle East is being torn apart by conflicting segments of a belief systems and tension is growing in the West as similar conflicts have found their way into democratic political systems. The wisdom of the founding fathers of the United States to keep church and state separate has never been more poignant. The vital separation of religion and the civil workings of state are slowly being narrowed by subtle and persistent manipulative political efforts.

Having served as a security guard in front of abortion clinics to keep protesters from blocking driveways or causing a traffic accident, I have been confronted directly by individuals driven by frustration to threaten me and others associated with the clinic. These programmed believers deflected any attempt to reason with them and a mutual understanding was impossible. Every argument was met with a display of their holy book or a rush to recite the rosary. The gates to any mutual understanding were closed tight as the protestors shouted and prayed and attempted to block or dissuade abortion clients. They ruthlessly attempted to convert drivers by stuffing pamphlets into their car windows as they passed, or by shouting at anyone who would listen. For them there was only their faith and no outside influence would ever dissuade them or get their attention. Their minds were closed.

Disturbed by the growing encroachment of religion into our politics I occasionally watch one of the many religious television programs. A recent program got my attention as the host and a guest author seemed to be revisiting the Scopes trials. The author was highlighting his recent book, and describing how any secular argument against God must draw on reality, (which, according to him, was obviously created by God), therefore, because the unbelievers were arguing against God by pointing out God's creations, any secular argument was moot.

For the author of the religious book under discussion, everything was proof of God and he had developed a favorite defense when arguing with nonbelievers. He used a question he claimed always put an atheist back their heels.

The question was; If Christianity is true would you become a Christian? He felt that any hesitation in answering was proof that he had made an irrevocable point that could not be countered.

In fact, he was just bringing up the old paradox about liars always telling the truth. The hesitation to answer he rejoiced in, was not a proof but the realization by those being questioned, that the question was a trick and had no answer. The correct response should have been a responding question. "If Christianity is false would you be an atheist?"

Attempting to support his guest's arguments, the host offered his own recent epiphany for the existence of God by relating how an earthmover at a construction site was also proof of a creator. His argument was that, even moving dirt requires a bulldozer, therefore, God must exist.

The extremes of religious TV programming and their increasing numbers may be evidence for a growing frustration by the devout. The silliest of early indoctrination programs for children is evident in cartoons about moral behavior by carrots, animals and robed biblical characters.

It took me a long time to extricate myself from my own lengthy indoctrination and leave the beliefs of my family behind, but now that I am a part of the real world again, aware of my responsibilities to my community, my country, humanity, and the planet, I am not about to return to the world of the disengaged.

Far more important than the ten commandments and a promised free pass to an after life, are the mandates imposed by 3.5 billion years of chance evolution bringing mankind to his advanced state of awareness. To ensure the future of our planet and our species we must think for ourselves, explore and learn, be of good council and, be good stewards.

As a species, we are unique in the solar system and may be one of only a few intelligent species in our corner of the Milky Way Galaxy. Being the result of billions of years of chance evolution and a single success among millions of failed attempts assigns great responsibilities, responsibilities far greater than being a servant to an invisible supreme being. We are discovering our true past in our genes, in the bones and artifacts of our ancestors, and in the stars and planets around us. These discoveries now provide the most reliable guides for our planets future and stand against the denial and distortions of those defending religious indoctrinations.

12

WEIGHING IN ON THE FERMI - HART PARADOX

Years before we began to confirm the existence of planets around other stars, we began looking for evidence of extraterrestrial life. Assuming that an advanced life form would be following emergent patterns similar to our own, and would be using similar technologies, we began to listen to the nearby stars hoping to intercept their version of "Saturday Night Live". We have been listening for a long time, but the rest of the universe remains silent.

During this period, we have convinced ourselves that it is unreasonable that among the billions of stars in our galaxy, our sun would be the only star with a planet supporting life and humans the only technologically advanced life form. For most individuals the question is unimportant and they are satisfied with science fiction stories and entertaining monster movies as an answer. Scientists, on the other hand, have taken the paradox seriously and, like scientists always do, began developing formulae based on probability and statistics to find answers. Without evidence, their efforts only highlighted the paradox. As Fermi shouted out in frustration while having dinner with Teller and other scientists; "Where are they"?

We now know that most stars have planets, that the sun is a large yellow star, not an ordinary small star, and that we have been missing the most common stars, red dwarfs, which often have multiple planets. We have also discovered that some of the three hundred moons in our solar system have conditions suitable for the development of life, and the statistical guesses made by Fermi, Teller, and Hart were far short of the hundreds of millions of possibilities projected by our recent discoveries. Now we have no choice but to ask again. "Where are they?"

The paradox, having been deepened, has prompted many speculative answers. Among them that;

1. Technologically advanced life forms destroy themselves on a regular basis before they have much to say to their neighbors.
2. That living forms rarely evolve to a level where technology emerges.
3. That advanced life forms choose to keep quiet out of paranoia.
4. That they are stealthy hunters choosing not to reveal themselves
5. That they use communication methods we can't monitor.
6. That they have been monitoring us and, finding us unsuitable are avoiding us.

and on and on....

I choose to respond only to the proposition that we are not hearing the echoes of other advanced civilizations because very few life forms advance to an aware state allowing an investigative approach to understanding their surroundings by using tools to enhance their senses. Our observations and comparisons of the development of hundreds of extant and extinct phyla and species makes it clear that; the fragile chemical messages formed from ordinary matter to optimize the continuance of life by directing increases in complexity, is difficult to sustain.

Form and function emerge suited to fit changing environments and persevere, not because the better are chosen, but because the less suitable are eliminated. The process only works if there is a large number of living options to be tested against rapidly changing conditions, and only if the testing is subtle. Gentle nudges to a more survivable form create a continuing evolution. Gross changes create mass extinctions and force evolution to start over.

The form and function of living forms are not adjusted directly. Only the chemical language that directs their emergence is adjusted as form and function are tested against the environment. The living form struggles to survive and reproduce in the short term, the gene struggles to survive and reproduce in the long term, and both are prompted solely by natural causes similar to sea foam forming on the sand of an ocean beach. The process of evolution follows natural paths similar to the combining patterns of atoms and molecules as they create successive levels of increasing complexity. We have agreed that matter follows the same formative patterns throughout the universe and assume that life follows similar natural formative patterns in developing adaptive forms and functions.

Reasoning abilities are needed to transform minerals and bone into tools and to create the technology needed for cooperative advanced civilizations to develop. Genes are effective in regulating the simplest of living processes. However, until ways to interface with and modify the environment emerge in bacterial forms, genes remain ineffective in producing more complexity.

The random emergent patterns needed are genetic accidents that improve sensitivities to temperature, chemical conditions and internal conditions, and each random adjustment that improves survivability is passed on and over time these advances in microbial awareness become cumulative and lead to parallel developments in form and function.

The partnership advantage between the development of complexity in form with advances in levels of awareness leads to advanced sensory and cognitive abilities. This parallel progression is evident in all life and is so interdependent that it seems natural to assume extra terrestrial life would follow the same parallel pattern of development and advance to a maximum optimization of awareness until, like humans, they assume control of adaptive evolution and override survival selections made through natural selection.

Advanced states of awareness similar to our own require advanced technology and advanced communicative skills, but not necessarily a pace or scope of awareness similar to humans. The many planets that orbit red or brown dwarf stars can have years lasting less than a week, and days lasting only a few hours. We may be attempting to communicate with an advanced life form with a pace of awareness faster than a humming bird or slower than a snail.

Time is not the only thing that is relative in our strange universe. The extreme numbers of planets and moons in our galaxy and hundreds of billions of other galaxies makes it unreasonable that we are alone but also makes it unreasonable that advanced awareness elsewhere will have similar perspectives. If we are to hope for success, we need to look for other reasons for our failure to find advanced life and develop search methods other than scanning the electromagnetic spectrum. The numbers are too large to ignore, the elements needed for life too common, and the planets and moons around other stars arranged in ways suitable for life too common for us to be alone, but ..... Where are they?

13

The Little Hiss

As a child, I played with rockets and cars powered by small CO_2 cartridges. The cartridge was placed in the toy and a spring-loaded needle was used to puncture a plug in the end of the cartridge. When I pulled the plunger back and released it, the needle opened a small hole in the cartridge and the compressed CO_2 inside was released with a hiss and a cloud of condensate was created as the potential energy stored in the compressed gas was released.

According to current scientific theory, our universe started in a very similar way. In spite of having been poorly named "The Big Bang", the energy that created the universe, like the energy that propelled my toys, was held in a small compressed state, (not big), and was released as a hiss, (not a Bang). We can still detect the hiss as microwave background radiation. Simply tune your TV to an empty channel and turn the volume up to hear the actual hiss.

My toys were powered by the release of potential energy. So is the universe. When the energy that powered my toys was released it created movement and a condensate, (dry Ice). When the energy that became the universe was released, it created movement, (expansion), and energy condensates. The energy release that powered my toys was inertial energy converted from the potential energy stored in the cartridge when the Co2 was compressed. The universe began as a compressed singularity and when its stored energy was released, both movement, (time), and several forms of energy were created. Initially there was a great rush, called the inflationary period. After the initial rush, momentum took over. The last term, "Tuv", in Einstein's basic field equation for general relativity is a stress energy momentum tensor describing this momentum. All of the energy released from the singularity at the birth of the universe and all of the energy released from my Co2 cartridges is still with us. It hasn't disappeared, it has only been transformed and occupies a greater space.

We have now identified some of the basic condensate energies resulting from the creative release at the beginning of the universe including;

- gravity
- electromagnetism
- the strong force
- the week forces
- and a special energy condensate called matter

We have yet to identify the most prevalent energy causing the universe to expand at an accelerated rate but, the acceleration theory is based on the assumption that the speed of light, (our measure of time), has been constant as the universe expanded. If light speed is not a constant, we may be drawing a false conclusion.

When the Co2 escaped from my toy cartridge, it created a temporary energy deficit and a dry ice condensate that existed until the heat energy, (created when the Co2 was compressed), dissipated. As the deficit was resolved by drawing in heat energy to overcome the cold period deficit, sublimation occurred, the dry ice was converted directly from a solid state to a gaseous state and the deficit was resolved by converting the re-gathered heat energy into molecular motion in the re-formed gaseous Co2. A similar energy deficit seems to have occurred when the Universe hissed into existence and, a short 300,000 years after the Universe escaped from its singularity, primal condensates emerged as electrons, photons, quarks and gluons but, unlike my toy cartridge, the energy deficit has not been resolved.

This continuing energy deficit may help explain the entanglements of quarks in glass energy condensates and how mass and gravity became topological features of space. My interpretation of the creative event may not be a fully developed theory but, hopefully, will prompt others to search using new perspectives.

14

WHEN THE STARS SPEAK

It had been a bummer of a day. Jake had not only flunked his astronomy exam, he had also lost his girl friend when he claimed to be smarter than her. In Jake's mind, none of this was his fault and, feeling sorry for himself, he purchased a six-pack of beer and found a secluded park bench near the campus rose garden. As he opened his third beer, he noticed a single bright star in the darkening sky and impulsively started to recite a poem he had learned as a child.

"Star light star bright, first star I see tonight, I wish I may I wish I might................."

Before he could finish, Jake was rudely interrupted by a deep commanding voice…. "You don't really think stars grant wishes… Do You?"

Jake looked around.

There was no one there.

"Up here", the voice said. "You started the conversation now finish it!"

Jake looked back at the star and remained silent.

"Yeah, smart ass, its me, I can tell your enjoying a few of my photons, so what the hell do you want?"

Jake muttered a few unintelligible words, his gaze transfixed on the brightening star.

"Speak up man. I can't hear you!"

Jake had always thought of stars as friendly twinkling lights, but this star was an angry SOB.

"I was just wishing," Jake replied. "It didn't mean anything."

"You've got that right." The star replied. "You think of yourself as something important and me as a tiny bright spot in the sky, but you have it backwards. You are the tiny spot and I am an enormous nuclear furnace. I was shining long before your solar system formed and am blasting radiation in all directions, up, down, to the sides and behind me. The tiny bit of light you see is an infinitesimally small part of all my radiation and there are millions of life forms on your tiny planet and on millions of other planets looking at me from all directions, and yet only you and your species assume that since you can see me, you own me. You depend on a star you call Sun, which in my opinion is a really dumb name for the energy that allows you to live, and you need to consider that you are made of stuff created in the belly of long dead stars. There are only a few billion of you humans. There are millions of billions of stars, more stars than all the grains of sand on your planet, and yet you claim the right to pick out one of us and get wishes granted. You really need to get a grip on reality!"

Jake sat very still. The sun was just beginning to peak over the campus buildings behind him. The six-pack was empty and the stars were gone. Jake's inflated ego had been flattened. He went to his astronomy professor and begged to take the test again. His professor smiled and shook his head, No. Jake then found his girl friend and asked for forgiveness, she told him to get lost. Later that night Jake crossed the campus after his last class keeping one eye closed. He glanced quickly at the wishing star but kept his mouth shut and quickly looked away.

Before we were blinded by city lights and before we viewed the stars through telescopes, man marveled at the star filled sky and imagined Gods revealing themselves outlined by dots of light. Before the stars began to speak to us through science they were benevolent lights in the heaven put there to guide and aid humans. Now that we understand their language, stars are revealing themselves as massive balls of fire, exploding with forces that echo throughout the universe, collapsing into black holes, revealing their ages, showing off their planets, and asking us to put our self-aggrandizing egos aside. And yet we, like Jake, close our minds and ignore our new perspectives and the mandates imposed by being a part of the marvelous spectacle that surrounds us. We are born of the dust of exploding stars and gathered by gravity and chemistry to become the vessel that provides nature with a mirror in which she can examine herself. We are indeed privileged, but we are not free to ignore what the stars are telling us. They have waited a very long time for us to listen.

15

ARE BLACK HOLES HOLLOW?

Setting aside electrostatic attractions and magnetism, there remains an attractive force that causes an apple to fall toward the center of the Earth and holds our planet in a nearly circular path as it orbits the Sun at 66,000 miles per hour. We have described this mysterious force in detail but haven't been able to explain it. To measure this force and to create a unit we can use to compare the attractive forces generated by various concentrations of matter, we have created a mental construct we call mass.

We establish the value of our construct by observing relative motions caused by the attractive forces of various massive objects. A small object accelerates toward the earth at 32 feet per second, per second, and orbital velocities and distances are used to estimate the mass of large orbiting objects like planets and moons. Until recently, this relationship between concentrations of matter and gravity seemed absolute, without matter, there could be no gravity and matter could not exist without producing gravity. Einstein explained the relationship by assuming that mass warps the space surrounding it and further that it warps time.

As if this wasn't confusing enough, we extended our concept of mass to any accelerating force resulting in mass becoming, not a weight, but a potential energy dependent upon relative position and relative motion.

When objects are separated, or are in contact, we can adjust our concept of mass from one described by motion to one described by energy;

A *potential energy* created by separation and/or motion.
Or,
A *pressure energy* created by contact without relative motion, (*arrested motion*)

Our new descriptions are equivalent to the measurements and descriptions currently used to describe mass. Potential and pressure are only an alternate view of the same mysterious force tending to draw everything in the universe back together into one big, or very small, clump.

Recently we used our advanced ability to calculate orbits and measure mass to asses the revolutions of whole galaxies and discovered that either our math or our understanding of the relationship between gravity and mass was wrong. Unwilling to give up long held assumptions we concluded that there must be an enormous amount of invisible mass creating enough gravity to explain the observed rotational anomalies, and "dark matter" was added to our conceptual repertoire. An easier explanation would have been welcomed, but using new telescopes to observe the bending of light from distant objects, the presence of dark matter seems to be confirmed.

Have we described these effects accurately? Newton was the first to describe gravity in mathematical terms by using the calculus. He also developed formulae to describe gravitational effects under special conditions. One of these conditions was how gravity behaves on and in massive hollow spheres. On the surface of Newton's hollow sphere gravity attracts everything toward the sphere's center, but inside, (on its inner surface or anywhere else inside), gravity cancels out and becomes zero. Newton's formula for this condition is on the Internet.

Another curious Newtonian gravity effect is pressure. On the surface of a massive sphere, gravity attracts all things toward the sphere's center. Inside the sphere, matter near the surface is attracted by all of the matter below, extending to the other side, but deeper in the sphere, the matter above begins to have an upward attraction and there being less matter extending to the other side the attraction toward the center is decreased. The mass above and below continue to increase and decrease until we reach the center of the sphere where they are balanced and both gravity and pressure become zero.

From this perspective the highest pressure energy occurs, not at the center of the solid sphere, but at about 3/4 of the way from the surface to the center. This disparity from the assumption that gravity increases from the surface of a massive spherical, or disk shaped object, at a constant rate from the surface, or rim, to the center, raises questions about conditions at the center of planets, stars, and black holes. Einstein's *feldgleichengen der gravitation* has proven to be a better explanation by describing gravity as a relationship between mass and the shape of space but the relationship is not a variable. If it were a variable, galaxies might rotate as observed and black holes might be hollow. If black holes are a massive hollow shell with an outer event horizon and an inner event horizon the paradox of black hole evaporation can be resolved and, If the event horizons act as a matter/ anti matter separator with antimatter evaporating inward, they may act as a secure vault for antimatter. A breach in the event horizons, allowing matter and anti matter to mix, might also explain cosmic ray blasters and quasars.

16

DIGITAL CLOUD OR DIGITAL FOG?

The Digital age has transformed the way we communicate, the way we educate, the way we shop, the way we design, the way we build and the way we think. Digital information, (on or off) sequencing, is not new. Nature has been coding information this way since the Big Bang. Life is coded for duplication by chemical chains of (on or off) sequences and we think using (on or off) activations of complex electro-chemical neural pathways. What is new, are human tools that mimic nature's information storage and transmission methods.

At first these tools were as simple as beads on wires placed in an on or off position, switch boards with lighted arrays, and punch cards, (hole or no hole). Now, silicon chips using microscopic printing techniques can store and manipulate vast amounts of digital information and more importantly, can translate alpha-numeric languages into a digital language and back into alpha-numeric's that humans can understand. Our success as a species is credited to upright walking and an opposable thumb, but just as important was an articulate tongue. Our communicative skills were made possible by a combination of a voice box evolving back and down and an ability to manipulate lips tongue and air flow to produce a great variety of sounds. With sounds representing objects, (nouns), and sounds representing actions, (verbs), humans gained dominance in the biosphere by using their ability to communicate.

Most of human language history was oral, but as cooperative associations became larger, another method emerged and we began to use visual symbols. Early crude representations of animals and events drawn in the dirt and on cave walls were quickly associated with the sound symbols for the common object and written languages began to form.

At first, the visual symbol and the sound symbol for a common object were not directly associated. The advantage of combining these symbolic steps by having a sound for each visual symbol eventually connected a voice sounds to each visual symbol and for those who understood the code the marks in the dirt could be made to speak. Sound symbols scratched onto rocks made spoken words permanent.

Humans took their first step away from a direct sense and react relationship to the things in their environment when they began using symbolic sounds. They took their second step away from a direct cognizant connection with their environment when they began using visual symbols to represent sounds. Both steps created symbolic thinking. Recently we have created a way to augment both symbolic steps by using an intervening system that captures and allows the manipulation of symbolic representations in a language using a new form of logic, (digital).

Digital intervention has given us a new way to interface with reality and serves as an intermediary between memory, reason and conclusions denied us using the nuanced confusion of sound and visual symbols. This digital intermediary may be bringing us closer to an understanding of the reality around us, but it may also be separating us further from the answers we seek. The technical advantages of our new digital approach seem obvious but its greatest impact has been to shift the cooperative structures that form and sustain modern civilization from well informed to over informed, and from a small amount of relevant information to a large amount of irrelevant information. Until we fully understand this shift, our economies, our governments and our interpersonal societal and international relationships are adrift. Have we created a beneficial cloud of information or a fog?

17

INFINITIES, LARGE AND SMALL

Infinities are enigmas that hide in the formulae of theoretical physicists waiting to pop up near the end of lengthy calculations making their efforts meaningless. Infinities stymie our investigative efforts and poison our logic, yet we give them status by assigning them their own symbol and by giving them names like singularities and black holes. We give them status to make their embarrassing snubs appear to be logical conclusions.

My keyboard has no infinity symbol but we all know that it looks like a twisted circle creating an intersection of lines at its center. The symbol was supposedly chosen to point out the fact that when one reaches the end they are back at the beginning, an appropriate reprimand for mathematicians and physicists that have followed their symbols to a dead end. The endless pursuit leading to infinities seems to have become a game played by theorists that is lost whenever one wins.

There are many functions leading to infinities. Some take a diminishing path to reach the absurdity while others take an expansive path. Are there, then, both large and small infinities?

18

GRAVITY <=> ACCELERATION, PQ - QP = X, F(X) = 1/X

Until recently, we held these truths to be self evident,

- Gravity is the product of matter.
- Acceleration is the result of energy changing the velocity or position of matter,
- Time is a fourth dimension represented by a constant (t),
- The speed of light through space is a universal constant

We have also recently observed that,

- There appears to be a lot more gravity than can be the result of visible matter
- That there is an apparent lack of equivalence between centrifugal and centripetal forces in spinning galaxies
- That time becomes a variable in gravitational fields and during acceleration
- That light can be deflected and slowed by gravitational fields
- That the universe is expanding at an accelerating rate.

We now suspect,

- That a concept we have long denied may actually exist as an energy field far more prevalent than visible matter,
- That Newton, Einstein and Heisenberg may have missed an essential piece of the puzzle
- That our investigative efforts to understand the accelerating expansion of the universe should be focused on dark matter and dark energy.

The dilemma our observations create, however, may not be the result of new mysterious forces or of new states of matter. Instead, we may be misinterpreting of our observations due to false assumptions. Scientific advances are the result of careful observations, carefully thought out theories, and many experiments to verify any new assumptions. These efforts build on each other over time, but this wouldn't be the first time long established scientific principles have been tripped up by a new observation, nor will it be the first time new observations are proven unreliable. The only way to tell if our past assumptions were wrong or if they were correct and we are misinterpreting the new evidence is to keep searching.

We can and should search for the newly implied dark matter and dark energy but we should also be looking for an error in our established measures of cause and effect at macro scales where (t), time, and (c), the speed of light, are determinant factors.

Time, *a mental construct*, is included in our equations as a quantitative constant, to make the equations real. The speed of light is used as a constant to make our equations balance but, what if the speed of light is a universal variable dependent on the expanded state of the universe and not a constant? If the speed of light is a variable, new perspectives and different conclusions result. If the speed of light is a universal variable dependent on the expanded state of the universe, mysterious dark energy may disappear and mysterious dark matter be explained.

19

PERSISTENT PERCEPTIONS

As you view and interpret the first word of this essay it slips into the past while you view and interpret the next word and so on until you get to a little dot that tells you to recall all the words in the sentence and form them into a connected whole. This synaptic exercise is then stored and compared to other stored synaptic groupings as a part of an expanding record of all your perceptions, everything you have ever experienced. We often confuse the duration of what we call the *present* with a collection of recently recorded perceptions, (the time it has taken to read this far in this essay for example). The duration of the present is arbitrarily assigned by every living thing with an active awareness but, in reality, the present is much more fleeting. What gives our thoughts continuity is the persistence of our perceptions as sets of synaptic captures give our awareness a temporary permanence as perceptions are carried forward.

The material world around us, and of which our bodies, sensory organs and synaptic organs are made, also posses this ability to persist as a thin event horizon moves forward in tiny flickering jumps. Combining these flickering moments gives us the illusion of permanence and a wider present.

In the real world change occurs much more rapidly than our

perceptions, and although not obvious from our limited perspective, alters everything around us at a dizzying pace. We have begun to recognize this discrepancy between our perceptive ability and the flickering of events as a quantum effect and struggle to deal with the discrepancy by attempting to reconcile quantum effects with relativity. The effort to join quantum reality with a relative world is beyond most of us conceptually, but the concept that we live primarily in the immediate past and control the future only by fleeting synaptic choices and recorded perceptions is easier to comprehend.

As you walk from one room to another, you put one foot in front of the other with the direction of travel dictated by a decision made in the recent past. As you transfer your weight from your left foot to your right while you walk, your last step is now in the past and your next step in the future. We overcome the need to provide synaptic oversight to micro moments and every step by creating the illusion of an extended present moment lasting from several steps at a time to the entire trip into the next room. The practical world of our existence is created by our ability to combine many flickering present moments into a single perception. The past is gone forever as soon as the electrons in an atom change position or the earth turns a fraction of an inch. The future state and position of all things would be predictable if, the number of atoms, objects and interactive forces were not infinite and, if matter having become animate, (life), hadn't developed awareness and choice. We, and all living things, perceive permanence only because our perceptions, like the reading of this essay, are recorded and move together as a unit into the next fleeting moment of reality.

20

VINDICATION

My interest in science and technology began in my youth and has persisted into my declining years. I have lived during a period of significant scientific discoveries and great technological advances. Memories of a time before television and computers have given me an appreciation for the rapid advances made by science and our ability to understand and apply natural processes. Jumps in technology are etched in my psyche by events such as the sound barrier being broken by the Bell-X1 in 1947 and flying at more than twice the speed of sound as an Air Force pilot only twenty-five years later.

My early school years were filled with scientific textbooks in a constant state of revision as Einstein's theories overturned Newton's concepts and as quantum effects became accepted. Our relationship to the universe around us has changed drastically in the past few hundred years.

Following the advances of science and cosmology as they have delved ever deeper into the secrets of nature has been a great adventure, but it has also been humbling. Great minds introduce concepts and theories that I could only partially understand and I struggled to deal with mathematical formulas beyond my ability. I began to feel excluded from the joy of discovery and the secrets that geniuses of the day were sharing with their peers but then theory began to reach beyond experiment, experiments began to produce conflicting results, and my reverence for the geniuses began to diminish. As great mathematicians and visionaries presented nature's truths as closing arguments to the search for a theory of everything, a strange thing happened. Using new observational tools, our nearly perfected theories ran into an old nemesis; *Gravity*. Could it be that the long chain of theories and discoveries had weak links? Could it be that my own thought experiments into the nature of space, time, and matter, might be as valid as the thought experiments of the esteemed?

Retaining the *assumed* relationship between gravity and visible matter, we have been describing and measuring gravitational effects in detail since Einstein's theories replaced Newton's, but when careful measurements of the visible mass and strange rotation of galaxies revealed there wasn't enough visible mass to hold them together at their observed rates of spin, we couldn't argue with the math. The outer stars of galaxies were spinning at the same rate as the inner core and the galaxies could only remain intact if there was much more gravity than the observable mass of the galaxy could provide. Unwilling to give up the dependent connection between mass and gravity scientists now search for invisible matter in the form of invisible dark atomic particles and I feel vindicated.

The great minds of our time never claimed omnificence, but I assumed it. Now that I know it isn't true, I can continue my own thought experiments and continue to write without feeling quite so stupid.

21

A PRIVILEGED ACCESS TO REALITY

We have assumed a privileged place in the universe since we became self-aware and appointed ourselves to a divine position. Unfortunately, our inflated self-importance has clouded a realistic assessment of our actual status, and for most of our history has separated us from the reality around us. We still assume a special access to reality that is all-inclusive and of a different kind, but using scientific tools of discovery; we are beginning to penetrate the egoistic fog obscuring a real universe that is complex beyond our ability to measure and more amazing than our best collection of miracles. The view we get through our scientific tools and predictive manipulations is however, still without a reasoned assessment, it remains a jumble of astronomical radiations and movements, genetic puzzles, quantum surprises and biological mysteries.

As we step down from our self-appointed throne we are learning that our assumption of a privileged access to reality, through our senses alone, is unrealistic. We are also learning that the only effective way we can deal with the blaze of the electromagnetic spectrum, unexplained gravitational anomalies, and strange atomic structures, is to create visual and mathematical analogies capable of being processed by synaptic nets molded to accommodate much simpler biological sensory inputs. We are explaining the reality exposed by our sensory enhancing methods using visual cartoons, digital reductions, and symbolic manipulations. We are incapable of a total interaction with reality because we, like all other living forms, have had our attributes of awareness naturally selected by random *best-fit* survival tests that have resulted in our current limited aware state.

The wide view of reality we are opening is threatening because we are not equipped to deal with the broad perspectives and the unfamiliar information from sources outside the range of unenhanced sensory information. Recognizing this disconnect from reality has opened several doors, some beneficial and some detrimental.

Our escape from a self-assumed privileged ability to access reality through our senses alone, or from revelations from a higher power, has led to an organized examination of the reality around us with technological benefits. It is also slowly leading us to a realization that all living forms possess some level of awareness, each with their own advantage, many beyond human abilities. This realization is fostering a greater empathy and appreciation for all life.

On the detrimental side, by pointing out our disconnect from reality and by challenging the idea of privileged access, a door has opened for those threatened by such notions to reach out to others feeling displaced and using rhetoric to radicalize their discomfort, are turning their doubts into narrow and extreme religious views. The fear and discomfort fostered by a sense of no longer belonging to or being able to understanding the much larger universe exposed by science, is fostering attempts to return to a pre-scientific era using revolt and violence, but we cannot undo a civilization built on new understandings without destroying ourselves.

We are the product of a universe with more stars than there are grains of sand on all the beaches on earth, and of matter becoming animate and alive through natural processes. We have become what we are through the same random process that have produced every other living thing from microbes to mollusks to mice, each one aware of the reality around it in its own special way. Moral codes formed in the fog of our egocentric past are still relevant but no longer sufficient. We now control, and must use wisely the power that is the energy of stars and our the power to control the future of life through genetic manipulations. If we turn back now civilization will collapse. If we wish to survive the imbalance between the powers we have corralled by observing nature, and our inability to act in accord with natural processes, we need to recognize new mandates imposed by our advanced evolved state.

We can survive our petty perspectives and egocentric views only by continuing to explore and learn, by recognizing the sanctity of language, and by being good stewards. All living things have access to reality and our access is not privileged nor is it complete. It is only more encompassing.

22

FINDING ORDER IN A PUSH-PULL UNIVERSE

More than any others, two men influenced our modern perspectives when they described the motions going on around us in mathematical terms. One, on a break from the formal curriculum of an English University to escape the plague, pondered and described the motion of falling objects. The other, influenced by the relative motion of passing trains, pondered and described the effect of objects moving at high speeds relative to one another.

Both Newton and Einstein used mathematics to measure and compare the motions they described, and both have had a profound influence on advances in science, technology, and human perspectives. Both men described and created comparative systems for the motions they studied but neither explained them. Explaining what they described waits on further discoveries and the serendipitous conjunction with other alert minds.

Gravity goes unnoticed because it is always there, seemingly immutable and innate. Newton asked why and came up with the idea that all objects with mass attract each other with a force dependent on their combined mass and the square of the distance between them. The solution he came up with to answer a *why* question, turned out to be only a *how* answer. We still don't know *why*. If we did, we could explain the gravitational effects holding the galaxies together and determining their spin when there is no visible mass to explain it. We have named the unknown cause of the anomaly we observe *dark matter*, assuming it is some kind of hidden mass, but all we really know is that we are observing an unexplained gravitational centrifugal effect. Newton also described but did not explain inertia. Matter at rest tends to remain at rest and matter in motion tends to remain in motion. His observation, *like all things fall down*, are DUH! type statements. We all know things fall down without having it pointed out. What Newton added was not an explanation for gravity. What he added was the concept of mass as the cause and a mathematical way to measure it. Newton also related gravity and inertia using his mathematical concept, *mass,* and at first glance gravity and inertia appear to be related because they both are measured using mass, but they are quite different.

Before Einstein, one could hypothetically carry gravitational pulls and inertial pushes to infinity, but Albert put the brakes on by exposing a speed limit, (the speed of light). Extending his relative speed experiences on trains to the motion of micro and macro objects, he exposed a universe of endless interacting motions exchanging their inertial energies through gravitational and physical pushes, pulls and collisions. He also pointed out that changes in inertial mass at extreme velocities convert mass to energy and vice versa. Even duration (time) becomes a variable in this concept but it is all still a description, not an explanation.

Two other interactive forces play on the same stage as gravity; (inertia and buoyancy). If we fill a hollow Newtonian sphere with water it is no longer hollow but if we follow the reasoning above there is a zero gravity state at the center of our water filled sphere where there is an equal amount of mass in all directions at the center.

Now place a cork in the water on the surface of our sphere. In a lake, it will bob to the surface because its buoyancy is opposite to gravity. If, however, we place the cork near the center of a massive water filled sphere, will it bob to a weightless point at the center? A similar easy experiment to illustrate the relationship between inertia and buoyancy is to tie a helium filled party balloon inside a car so it is free to move back and forth. Close the windows and turn off the heater fan. Step on the accelerator. Which way does the balloon lean? Now step on the brake. Aside from being similar, inertia, gravity and buoyancy are not identical, just strange cousins. Is there a micro limit below which gravity doesn't exist? Can buoyancy partially offset inertia? If matter has an energy density less than the energy density of space, can buoyant energy explain gravity?

23

A VERY DANGEROUS QUESTION

Nothing separates us more than ingrained belief systems. Communal beliefs are generally acquired and embedded through repetitive indoctrination. Most are a carry over from ancient attempts to explain and order human affairs and have been made permanent by writings declared sacred and are sustained by ritual. Belief systems expand their membership through evangelism or conquest and develop internal control mechanisms, some despotic and some democratic. Members of established belief based organization hold their basic tenants of faith as god given and above question. Sensitive to contradiction, the faithful will defend their beliefs using force when necessary. Communities based on faith attribute their insight and special importance to a unique link to a god or gods and a special understanding of god's directives.

Protecting and sustaining religious organizations requires evangelical efforts, membership control, and a constant defense against other belief systems and any unacceptable societal changes. The primary societal changes that threaten the faithful today are the result of discoveries by science that make revelations and miracles suspect. When gods were more numerous and had many names, one could ask about one's beliefs as simply and safely as if inquiring about one's family or health without risking a defensive response. With the rise of monotheism however, and the idea that ordinary individuals could interact with a god without the intervention of a priestly mediator, faith based communities took on a separate status and became more vulnerable to manipulation, and more militant.

The question; "Do you believe in god?" is not a question you ask a stranger if you want to open a friendly conversation. It is easy to naively assume that everyone defines the word "believe" in the same way and that everyone knows and understands what you mean by god, but ask, "Do you believe in god", repeatedly at work, and you will attract the attention of a supervisor. Ask the question in a bar and you will find yourself sitting alone. Ask the question in a church and you will get as many questioning looks as positive answers. The reason for these responses is that there are as many gods as there are human believers, and belief is not a quantifiable term. Belief has many levels ranging from pretense to fanatical commitment. Even within well-established faith based communities schisms develop. Faith is a specific commitment to a codified set of behavioral directives dictated by a supreme being as interpreted by officials appointed or elected from within the faithful community.

When the question, "Do you believe in god?" is asked, more often than not, the answer involves elements that separate believers from non-believers and leads to definitions that identify those with questionable beliefs who might pose a threat. When differences are identified, mistrust is inevitable. Faith based conflicts are usually initiated, not by the secular or from within the community of the faithful, but by the priestly guides who see their power positions threatened.

Historical faith based conflicts are usually between competing faiths and radical belief systems, but as science feeds the growth of technology and exposes nature's secrets, the general population becomes more rational and faith based communities more defensive. Dangerous lines are being drawn between the faithful and the secular, between those who pray and those who do not. The ritual of prayer has become the control mechanism of choice used by modern despots who would exploit the faithful and turn them into weapons. For the faith manipulators, questions of faith have been replaced by the dictates of how and when to pray, and the better, and safer question to identify a believer from a non believer is, "Do you pray?"

24

A CIVILIZATION BUILT ON SAND

The familiar metaphor, that one should place the foundation of a house upon a rock rather than on sand, was intended as guidance for the early church, but the metaphor also applies to modern technology. Civilization began with the use of stone tools to shape other building materials. Civilization continued to advance with the use of bronze implements, took another step with the use of iron, and another with the use of mortar and concrete. Our inventive use of materials has brought us to our present state of roads and structures, all carefully placed on solid foundations.

Unfortunately, we found a use for sand that may be making our modern civilized world fragile. Silica, (sand), is the second most common element on Earth and we have used it as an abrasive, have turned it into glass, and combine it with many other elements to our advantage. Very recently, we discovered another use for this common element by using it in tiny chips to store and process information. Sand now allows us to create thinking self-directed machines. The stuff we let our children play in, and walk on next to ocean waves, now has a memory, lands airplanes, directs factory robots, guides bombs, directs the flow of electricity to our factories, monitors water distribution systems, keeps track of sales and inventory, is integral to Wall Street and is central to our communications and to navigation. We have replaced direct control of nearly every essential operation in our complex civilization with chips of sand and have become dependent on their reliability.

By taking our hands off the controls and reducing our attention to passive oversight, we have created a fragile interface between ourselves, and the systems that sustain us. In just a few decades, we have replaced a civilization resting on rock and steel with human hands on control levers, with a civilization dependent on sand based chips susceptible to gamma ray bursts, electromagnetic pulses and computer viruses.

When sand shifts under the foundation of a building the building sinks and the foundation cracks but, with the exception of an earthquake, a sinking building seldom has an effect on a neighboring building. Our new silica-chip-civilization however is networked into interlocking operations that communicate with individuals and each other and a crack in one foundation quickly spreads. Some of these cracks are accidental, as was the case of the great power grid failure in 2002, but some are intentional, as in the case of hacking into government employee records.

Even though shielded and isolated from the Internet, the security of the computer systems running the uranium enrichment program in Iran crashed when a powerful virus interrupted its programmed directives. Introduced internally by the insertion of a simple thumb drive, the virus put a Nation's nuclear program on hold. What makes this incident noteworthy is that it was intentional and that the destructive virus escaped the secure facility and subsequently infected thousands of critical operations in many other countries around the world. The implications of these incidents, and many others, should make it clear that our technologically based civilization, dependent on traces etched in silicon chips controlling communications and vital infrastructures, has moved the foundations of human survival onto soft ground. Ironically, it may be the inherent uncertainty of the quantum world that moves us back to a rock solid foundation.

25

MILITANT EVANGELISM

To be an evangelical one needs to be made privy to truths that are self-evident and beyond question. These absolute truths can be of a religious nature, can be ideological, or simply an elaborate lie designed to seduce large numbers of people to a cause. Nearly all evangelical efforts begin in a place of worship, in a conversation, or on social media and use emotional appeals. Evangelical efforts seduce by offering a reward or by tapping into latent paranoia or hatreds.

Evangelical efforts vary from the promotion of specific beliefs to instructions in proper methods of worship. Evangelism is also made evident in the promotion of a proper relationship between subjects and a king, a dictator, or more recently, in the promotion of proper economic systems for nation states. Evangelistic causes vary but they all start by identifying their audience as those who have not yet been informed, and those in opposition.

Evangelism can be,

- Passive; the Gideon's placing Bibles in hotel rooms,
- Direct; Mormon youth knocking on doors and leaving pamphlets, or

- Confrontational; as exemplified by the Crusades, and now by Isis.

There are sixty thousand evangelical Christians in the United States and over thirty thousand militant Muslim extremist now engaged in an effort to convert the world to their absolute truths and their narrow views. Militant evangelism takes place on small and large scales.

The bombing of abortion clinics and murder of abortion workers is extreme evangelism using elimination and intimidation to change society to fit within a circle of beliefs. The crusades used the recovery of holy relics as an excuse to eliminate an entire civilization of *non-believers*. Shiites kill Sunnis over minor differences, as have Catholics and Protestants: I personally have been attacked with a machete by a frustrated Baptist evangelist when I refused his gift of eternal life and his demand that I commit to his beliefs. Militant evangelism recruits and controls a following by promoting paranoia hate and fear using emotionally charged words, measured interpretations of ancient *holy* writings and by narrowly focusing educational curricula.

Evangelistic movements solicit followers and funding by appealing to basic tribal emotions, (once of value in a pre-civilized world, but now a dangerous vestige). Many potential recruits exist among those who feel displaced by technology and modern societies. These individuals are easy prey for any cause that reinforces their paranoia, justifies their hate, and simplifies the complex modern world. By substituting responsibilities to a god or religious leader for more complicated social responsibilities, recruiters simplify their lives. To create an army they empower new recruits by making them a part of a cause, give them a gun to make them feel powerful and surround them with other like-minded recruits.

Funding from evangelistic groups are also used to select and support political candidates with religious views intended to render secular government bodies impotent. Take evangelism to an extreme and you have a revolution. Even when one of these extreme militant evangelistic movements succeeds, it soon splinters and another extreme view is established.

Those who hope to save us from ourselves may be the most dangerous among us.

26

SHAPING SOCIETY

Societies continually adapt as their customs and laws adjust to fit changing conditions. As societies change, the balance of each individual's commitment to their society also changes. Societies are defined by their history, their geography, their economics, and by the commitment of its citizens to the ideas that distinguish it as a separate state.

A stable society is composed of individuals who feel secure, have their basic needs met, and voluntarily comply with the norms of behavior defining the society's moral and lawful essence. A stable society is one in which the self-interests of its citizens are balanced with their commitment to their society's well being, a balance between *Me* and *We* concepts.

Societies composed of dissatisfied and disenfranchised individuals are susceptible to internal and external disruptions. Societies with strict rules of behavior that suppress the *Me* in favor of the *We* in the *Me/We* equation, can appear stable but are potentially self destructive. Societies that overly emphasize the *Me*, while disregarding the importance of the *We* in the *Me/We* equation are also subject to failure.

Balancing logical commitments to ones society with reasonable commitments to self should be the primary goal of all societies.

Unfortunately, many leaders put personal and ideological goals ahead of the real responsibilities of their office and create unnecessary stresses within their governed states. Narrow goals that disenfranchise and create societal anxieties can result from unwarranted commitments to a religious belief, a misunderstanding of the proper relationship between a citizen and society, or from a lack of economic understanding.

Religion has competed with societal leaders for the allegiance of tribal members and citizens for millennium. Witch doctors and priests have vied for power with tribal leaders and kings, Popes have vied for power with emperors, and democratic societies appease religious leaders with exemptions and special privileges. The struggle continues as religious leaders in secular countries attempt to influence and infiltrate legislative and judicial actions while self proclaimed religious states, like Isis, wage war on all secular arrangements.

John Locke's admonitions, in his essay concerning toleration (1667), that church and state need to be kept separate for both the good of the state and the protection of religion were acknowledged by the founding fathers of the United States and subsequently adopted by most other Western Countries, but the struggle continues. Turkey's recent (2016) coup attempt is an example, (a struggle between secular and Islamic ruling principles).

Religion competes with secular rule by adding a third term to the *Me/We* equation, by requiring an allegiance to ephemeral beings and the rules of conduct derived from their edicts. A citizen of a state where secular laws prevail but religion competes, must choose to be secular, or religious, or both, and if they choose both, must balance two allegiances. The shifting balance between a citizen's concerns about their individual rights, their obligations to god, and their obligations as a member of a society, is an unstable determinant in any society.

Religion, when in control, is a poor master in modern societies because It emphasizes life as transitory, societal obligations as secondary, and religious obligations as primary. Theocratic societies control and direct their citizens by emphasizing the individual's relationship, not to the state, but to an *after death* existence and by supplementing god's ultimate judgments with contemporary standards of behavior that subjugate the *Me* in the *Me/We* equation to an unimportant temporary state.

Economic systems also shape societies by influencing the essential *Me/We* balance not only in their macro applications but by societal leaders influencing individual attitudes with intended and unintended incentives and disincentives. Macro economic systems are, in part,, defined by their *Me We* emphasis. Capitalism emphasizing *Me* incentives, and communism emphasizing *We* incentives. One, emphasizing profit, one emphasizing sharing. The problem with both systems is an inherent tendency for abuse and the resulting need for oversight creating restraints, which in turn creates a second opportunity for abuse.

There is no economic system immune to greed. Paradoxically, capitalism recognizes greed as both a proper incentive to create wealth, and a crime. Another misleading assumption is that certain forms of government are only compatible with certain economic systems, democracy with capitalism for example. In fact, how one chooses ones leaders and to what degree one allows citizen participation in government has little to do with a society's choice of an economic system. Even the most despotic regimes can be made compatible with the most liberal of economic systems.

Over or under control of economic incentives for individuals or businesses also affects the essential *Me/We* balance and societal stability. Incentives for innovation and reward need to be balanced with an acceptable distribution of wealth for any society to remain stable.

Beyond the factual conditions that shape societies there are ephemeral influences resulting from image and perception. Societies, like individuals, develop a persona based on their proclamations and actions, and, like individuals, are subject to prejudice and praise from other societies. Societies are also affected by changing internal perceptions when threats from outside or inside create social anxieties, economic imbalances, or when individuals within a society draw attention to themselves.

In the United States, the *Me/We* equation has recently swung to the *Me* side as economic disparities have become apparent and as anxiety has increased due to the threat of terrorism. In this imbalanced state attention to individual rights and individual security has become predominant, tipping the scales away from individual responsibilities and the security of the State.

Privacy and the right to bear arms has temporarily overwhelmed our sense of pride and unity in favor of a *Me* first attitude. Ideological gridlock threatens our democracy as internal economic disparities threaten our capitalistic approach to economics and our convoluted election process exacerbates the situation. The *Me* side of the *Me/We* equation that sustains all societies is unpredictable and emotional. The *We* side of the equation is based on agreement, perception, and commitment.

The most effective way to restore pride and commitment is to go to war. Unfortunately, this is not a reasonable option for any sane society in today's world of advanced weaponry but the option remains a desperate end game for dangerously despotic governments.

The best solution for the US is to separate ideological commitments, religious and political, from governmental activities, return to solving differences through reasoned compromises and by promoting less destructive talk-radio and more pride in the free press.

Attitude is important and candidates for office pointing out problems and proposing solutions can promote pride or create hate, fear and disillusionment.

A well-constructed societal assessment equation that measures the *Me/We* balance of a society would be invaluable as a predictor of societal stability, and could record how societies are shaped and how they change over time.

27

SOME AMAZING THINGS ABOUT LANGUAGE

While I was having a cup of coffee at Starbucks recently a young man next to me carefully marked his place in the book he was reading and commented that the book was amazing and that it had saved him. Trying to be kind, but at the same time attempting to avoid what was probably going to be a self indulgent disclosure of some tragic event in his life, I made a noncommittal response. Unfortunately, he was not put off, and as he told me about his past drug addiction and jail time he pushed the book he had been reading closer and I could see it was a Bible. He finished his story by commenting again on the amazing power of the written word and waited for me to open the door for his chance to save my soul.

I respect others perspectives and religious beliefs, but having been subjected to hundreds of pitches to join everything from multi-level marketing efforts promising quick riches, to evangelical recruitments promising eternal life, and have become dubious about conversations like the one the young man was trying to initiate. The amazing power of the word he was talking about was the word of God, but the actual words he was referring to were the result of writings by many individuals spread over six hundred years and translated from Greek and Aramaic into hundreds of other languages over nearly two centuries. The young man obviously found the book amazing. For him, it was a kind of magical code for behavior and salvation. I found the book amazing because it exemplified the human trick of coding meaningful speech sounds into visual symbols that no longer drifted away in the wind but lasted through time for future humans to scan with their eyes and translate into meaningful sounds. To me the miracle of the book he was pushing at me was not God talking to people, but the ability of humans to speak to other humans from the past, an amazing quality of all books.

As I pondered the amazing way plants, animals and humans communicate I remained silent and the young man, fearing he had lost the opportunity to fulfill his obligation to save my soul, repeated his opening statement loudly, hoping to interrupt my reverie. This time his loud and insistent announcement, that the word of god was amazing, got the attention of many others in Starbucks enjoying their own quiet conversations. He was now holding up the Bible and I was suddenly on center stage being threatened with a religious book by a militant evangelist.

"Admit it," he repeated. "The word of God is amazing."

Without thinking, I responded in an equally loud voice.

"What's amazing is that we can exchange complex ideas by vibrating the air between us"

There was a smattering of applause and laughter from other customers and the young man lowered his book, opened it to the marked place as if to retrieve the ultimate proof needed to force his religion on me, and went silent. He hesitated for a few moments then closed his magical book and left the coffee shop without vibrating any more air. Others have attacked me when I resisted their attempts to save my soul and in Vietnam, enemy soldiers were intent on killing me in defense of their perspectives. I hold no ill feelings toward those compelled to bring others into their circle of beliefs, and thank the young man at Starbucks for prompting me to examine language from a new perspective.

Nearly all living things communicate in some way or another. Some leave simple traces as visual clues or scents to be read later by others. Only humans are able to leave complex ideas and thoughts in electronic and graphic form for other humans to access at a later time.

This nifty trick makes civilization possible. A month from now I can return and read what I have just written and will be listening to my own voice from the past. I can also go to the library, walk between the racks and touch thousands of books that when opened allow authors long dead to share their wisdom and ideas. Amazing!

28

THE ME/WE EQUATION

Humans are social animals. We live in groups and depend upon one another for protection and physical and emotional support. Alone we aren't much of a match for nature, but as cooperative collections of individuals, humans rule.

Human success is due in large part to their language ability, which allows them to share complex ideas and imagine things no one has ever experienced. The nuance of evolving languages also absorbs orders of things hidden in Nature that go unnoticed, allows the naming of reoccurring events, increases predictive powers, and expands our scope of awareness. Beyond our articulate tongue, humans possess an opposable thumb and walk upright with hands free to make and use tools.

Humans name each other and the tribes to which they belong and differentiate between their personal needs and the needs of their tribe. Like all social animals living in mutually supportive groups, humans also establish relationships, take on leadership roles and establish rules. They also establish a relative balance between their personal needs and the needs of the tribe. This balance between the individual and the group can be represented as a *Me/We* equation, a ratio that is weighted and balanced by circumstance but is perpetuated and respected as a matter of perception and is sustained by the acquiescence of individuals within the group. Nearly all ancient societies were weighted heavily in favor of the *We* side of the equation. Ancient Individuals thought of themselves as subservient to the ordered societies around them sustained by myth and tyrannical rule. They marched in step without complaint. Order was imposed and maintained partly by fear of punishment but also by the feeling of security afforded by being a part of a greater and protective whole. The Roman Empire was expanded by force but persisted by being all-inclusive and assuring, when possible, that all of its citizens benefited by being under Roman rule. The *We* perception has been weighted heavily against the *Me* for most of human history. The *We* imbalance provided the glue that allowed empires to persist for centuries.

A few thousand years ago, however, a *Me* message began to take hold that, for good or ill, has tilted the equation toward the *Me* as being equally important, or more important than *We*, and humans have been struggling with this shifting balance ever since. One can establish numerical values and assign them to *Me/We* equations for periods of recent history and to kingdoms and nation states. In a *2Me/8We* equation, (the State is perceived by citizens as four times as important as the individual), which might describe communism under Stalin, for example.

I will leave the task of assigning values to various countries and various periods of history to others. I use the concept of a *Me/We* equation only as a critical thinking tool that can be used to shed light on many current organizations and political views.

Unfortunately, recent religions have added an additional variable to the equation. Monotheism has added *The Other Me*, a concept that, beyond one's responsibility to their tribe and themselves, there exists an inner *Me* called the soul that is responsible not to the tribe but to an all seeing God.

We now have *Me/We* equations like these:

OMe>Me/We, (The Other *Me* is greater than *Me/We*)
OMe<Me/We, (The Other *Me* is less important than *Me/We*)
OMe=Me/We, (The Other *Me* is equal in importance to *Me/We*)

An example might be a Christian living in Iran, (*8OME>6ME/4WE*): this individual places his religious beliefs above his responsibilities to self and the country he is living in but weighs his responsibility to self above his responsibility to Iran. Pushed into a corner he will follow his religious beliefs first, his personal needs second and put his country last. The *Me/We* equation can be applied to individuals to evaluate the strength of their allegiances to their religion, to their family, a political party, a terrorist group or a nation state, not as a clear measure of fact but as an illuminating exercise. .

29

INERTIA

Sometimes new descriptive terms and perspectives help explain difficult concepts, create alternative ways to view nature's processes, and create new and useful insights. Exploring these alternatives as an adventure of mind is as exciting as living adventures experienced by challenging nature directly. I have done both. I have felt the relentless tug of centrifugal force trying to drain the blood from my brain while making six "G" turns in high performance jet aircraft and seen luminous moss light the leaves of jungle plants on a dark night while surviving in the jungle. The universe is a wondrous place continually beckoning us to explore, asking us to use our advanced awareness to interpret her methods and movements and, I am continually seduced by her call.

Without having acquired the mathematical tools and education of a scientist, I can only search with logic and words for explanations that I find reasonable. I leave it to others to accept or refute my descriptions, and that is all they are, (descriptions). Nature presents many puzzles. Two of these are the potential energy of separation and the potential energy of motion. We know these better as gravity and inertia:

- Gravity is the potential energy resulting from the separation of objects with mass and is expressed as (acceleration), as gravity causes objects to rush toward each other, or as (pressure), when separation has been reduced to zero.
- Motion is the potential energy resulting from the acceleration of objects with mass and is expressed as (inertia).

The potential energy of motion, (inertia) and the potential energy of separation, (gravity) are only cousins but, in terms of their affects, they are identical. They express in the same way by producing heat when their potential energies are resolved by either pressure or collision.

The potential energy of separation converted to pressure inside stars raises temperatures to fifteen million degrees. At these temperatures the atoms making up the mass that created the energy potential to begin with, move at such speeds they overcome the electromagnetic forces separating them and fuse creating heavier elements and a small portion of their mass into even more energy.

At the atomic level, the result of the macro potential of separation (expressed as pressure) becomes the potential energy of the motion of atoms, which when they fuse, is expressed again as a conversion of mass to energy. The energy of macro mass separation becomes pressure, which becomes heat, which becomes micro motion, which converts mass to energy. An especially strange circle of events when one recognizes that matter, (mass), began as energy.

In the beginning, there was only energy, a singularity, the potential energy for expansion, a condensed state of energy opposite the energy of separation, (gravity). We haven't given this energy a name, or have we? Could it be the same energy we recently discovered, and call (dark)? When the energy of separation was first expressed, it created motion and a small part of it condensed to become matter. As matter was separated by the inertia of the big bang, the potential energies of motion and separation emerged. From the potential energy for expansion came matter and the potential energies of separation, (gravity), and (inertia). From these two emergent potential energies a cycle was created, a cycle from energy to mass to energy, a cycle between one infinity and another, one extremely small, one extremely large.

30

BETWEEN THE PAST AND THE FUTURE

We treat time as the fourth dimension and divide it into three categories; (Past, Present, and Future). We then cut it into many pieces we call durations, attoseconds, seconds, days, hours, light years, and many more. We use these as units of comparison to measure movement, to order events, to control our technology, and much of our lives. We treat time as a real thing, something we can save, use, and spend.

In a practical way, when focused on everyday activities, this perspective makes sense. Our scientific investigations, however, have created a very different view of time, one that goes well beyond the cyclical events of the earth spinning on its axis. Our tools of discovery have become ever more sophisticated and as we look further out into space and deeper into the structure of matter, time takes on new dimensions. Our macro discoveries have forced us to expand our durational units to cover periods many times larger than our practical units and our micro discoveries have forced us to chop our units of time into bits so small we need negative exponents to represent them.

We have also discovered that time is not a universal constant but a variable and find this so disconcerting that we still cling to old perspectives in many of the formulae we use to describe our new findings. The past now extends back beyond human history more than 13 billion years to the beginning of the universe, and the forecast future of the universe extends forward in an equally incomprehensible measure.

But what of the present? Is it also a variable? Is our *Now*, different from *Now* in other places, and how wide is the present? How much time exists between the past and the future? As I type, I tap a key in the present, but the last letter I typed is now in the past and the letter I am about to type is in the future. Is the present wide enough to encompass my entire typing session? Is it wider, or is it narrower?

Our awareness is a survival adaptation of sensory organs and synaptic functions that allow us to sense conditions and activities around us and to use our bodies to react appropriately. Reflexes are built-in for nearly instant reactions and Instincts are genetically hard-wired for reactions to more complex situations. But life has developed the additional synaptic survival trick of recording events from the past and carrying them forward into new situations for use as needed. When the brain is in a collection mode, it perceives activity as the present. When the brain is in a retrieval mode, it perceives the activity as the past. We sense only a small part of the reality around us because of the limitations of our sensory organs and sense time only within the limitations of our synaptic functions. Our pace of awareness determines our perception of time including both the duration of the present and the division between the present and the past.

Old perspectives are difficult to abandon, especially our concept of time as a real thing, a fourth dimension that is integral to our current perspectives, experiments, and discoveries. One of these inclusions involves the result of light being shined on a metal and finding that it emits electrons, and that the intensity of light doesn't affect the results but the wavelength of the light does. When the wavelength of light is increased the frequency of electron emissions decreases by an equal amount leading us to a constant. When wavelength and frequency are multiplied together the results are (3.0 x 10 to the 8th power meters per second), which happens to be the speed of light in a vacuum. We also discovered that an atom energized by light accepts the energy and subsequently gives it back in packaged amounts, which led us to postulate discrete electron orbital energy levels, the uncertainty principle, planck limits, quantum theory and more. After combining the gravitational constant, (G), the relativity constant, (c), and the quantum constant, (h) Max Planck arrived at a constant unit of length, the smallest of which, (one planck length) is the smallest distance a photon of light can travel at the speed of light in a vacuum. Anything smaller and units of time become dimensionless and our wavelength/frequency constant becomes an infinity that swallows the photon. Likewise a Planck unit of time, (5.4 x 10 to the negative 44th power seconds) is the smallest unit of time possible before physical laws fail and this limit is thought to have implications for quantum gravity, has led to string theory, quantum loop theory and other extreme scientific proposals.

Without credentials behind my name I am free to think outside the box and am forced to do so because of my lack of fluency in the language spoken inside the box, *advanced mathematics*. Without credentials, I risk only the amusement of those with a deeper understanding but am in no danger of being embarrassed and can offer the following five observations.

(1) Changing relative positions or conditions, or states, (Be they the spin of a galaxy, the orbit of an electron, the death of a star or the joining of atoms to become molecules), requires space

and energy but not time. Time is an illusion created by an observer making a synaptic record of motion or change that is a collage of quantum pulses occurring much to quickly to be observed individually.

(2) The Universe exists only as a succession of planck units. A strobe lighted reality with a flicker rate far above our ability to observe.

(3) Quantum uncertainty and action at a distance are the result of distortions as a jump is made to the next Planck unit.

(4) Information is passed from successive Planck units, one to the next in unique arrangements of discrete packets describing energy, matter, relative positions and relative motion. Small transposition errors occasionally cause information to fragment and recombine in a hierarchy of evolving complexities that describes the motions and combinations of energy and matter that we perceive as reality.

(5) Our persistent use of the term (t), (time), in descriptive formulae to guide our experiments and discovery efforts may be misleading, even when it is not explicit, (as in the speed of light) (d/t). Replacing (t) with comparative motions complicate our formulae but could lead us closer to reconciling gravity with the weak and strong forces and in reconciling relativity and quantum effects.

31

DESCRIBING GRAVITY

From a time when pre-humans were being bruised from falling out of trees, we have been acutely aware of the force that tugs on us from below. Regarded by early man as a spirit that pulled everything down to a lowest level, gravity pulled rain from the clouds and rocks and mud from the sides of mountains to bring them down from presumptive positions. Mayan myths included a world below the Earth's surface where other people walked upside down and where, when the sun set above, lighted the world for the people below. Many other ancient myths describe gravity and its effects. Gravity is a strange but familiar force. Modern science allows us to measure and describe gravitational effects but hasn't explained it.

Newton, on a sabbatical from his studies at Cambridge to escape the plague, developed a formalized description of gravity using a new math, (the calculus). To measure and predict gravity's effects, he also introduced the concept of mass as a measure of the amount and density of material needed to produce a gravitational or inertial effect. This mass/gravity equivalency remains the primary core of our physical descriptions. We use Newton's math to calculate basic orbits and aim artillery shells. Even after being called into question by Einstein, and after the recent discovery of gravity without any apparent causal mass, (dark matter), we continue to cling to our practical descriptions.

The Mayans postulated upside down people on the other side of a thin flat plane and were getting close to the truth. The persistent observation of a curved horizon hinted at a sphere and the Greeks, using shadow lengths taken at different latitudes, were able to calculate the size of the ball implied by the curved horizon. Persistent common sense opinions however, clung to a flat earth image for centuries until astronomy overtook astrology and adventurers sailed beyond the horizon and returned without falling off the edge. Only then did the full realization that we were stuck to a giant ball by a mysterious force become widely accepted.

Newton's new math established the mass of the earth and the earth's gravitational field as a measure of acceleration. Objects are pulled to earth from a height at an increasing rate of 32 feet per second/per second and he used this measure as a yardstick for measuring other massive objects and their orbits. In spite of his efforts, we still don't have an explanation, we only have a description, and sometimes the description results in some very strange concepts. Newton's math describes why we can accelerate a satellite to 18,000 miles per hour and have it continually fall around the Earth instead of curving back to the ground. It also allows us to calculate a speed, 25,000 miles per hour needed for the accumulated potential energy of a space probe's acceleration to exceed gravity's ability to pull it back. Strangely, Newton's math also describes a point of zero gravity at the center of the Earth and, if the world could be hollowed out, leaving only a thick crust, the surface people would still feel the effects of a reduced gravity but the Mayan upside-down people inside the hollow earth would be weightless. With the advent of quantum theory, gravity's description becomes even more convoluted. We smash basic particles of matter together looking for clues in cloud chambers but still don't have an explanation. All we have is a more complicated and confusing description. Is the proton the source of gravity? It has mass but so do some other sub atomic particles. Is gravity a by- product of the strong force? Is it created by the gluons buzzing around holding the quarks together inside the proton? We are surrounded by and made up of mostly empty space. Immense empty spaces exists inside atoms, between stars and between galaxies, but gravity from dark and luminous matter continues to tug on matter at great distances, continues to shape the Universe, and control how things move. We can describe gravity but still have no idea what it is. Maybe it is a spirit like our ancestors thought.

32

LIGHT SPEED ILLUSIONS

Not much progress has been made in the past ten years with the dark energy concept and as a practical matter, In spit of its cute title and strange nature, the general population could care less and it doesn't get discussed in grocery store check out lines. We have made some advances in describing dark matter but still have no idea what it is and, like dark energy, it remains dark.

Observing strange rotational anomalies in spiral galaxies, we concluded that the anomaly must be the result of an additional amount of gravity. We then used *in-the-box* thinking and concluded that if mass generates gravity, and additional gravity explains the phenomena, there must be more mass in and around the galaxies we can't see. Instead of leaving the anomaly open for other explanations we opted to allow our equations to remain viable and assumed a great deal of hidden matter. We call it dark, but it isn't dark, it is invisible, (not responsive to light).

What we are really looking for is a strange force causing galaxies to rotate in ways we didn't expect and can't explain, and it may or may not be caused by invisible matter. Not long after our discovery that a mysterious form of matter dominated the gravitational scene, we used a new telescope to look at light coming in from galaxies much further out than we had seen before and discovered that nearby galaxies were separating from each other faster than distant galaxies.

Ironically, Hubble, the astronomer our new telescope is named after, was the first to use the red shift of light almost a hundred years ago to reach the conclusion that the universe was expanding. Now a space telescope named in his honor was presenting us with evidence that it is not only expanding, it is accelerating. We were first presented with the dark matter paradox then with the accelerating expansion of the universe, and we face another gravitational challenge.

For a long time we thought that the observed increasing separation of galaxies was the result of the initial push by the big bang, (another cute but misleading title), and that the total mass of all matter in the Universe would slow it to a stop and then reverse. We forecast the age and future of the universe using a total mass estimate and I sometimes wonder if the guy who calculated the total mass was related to the guy who added up all the begets in the Old Testament to put creation at about six thousand BCE.

All of this scientific speculation about the expansion of the universe was based on observations but also on a lot of assumptions. Using the expansion rate, the amount of assumed gravity, and some heavy mathematics to formulate a history for our Universe reaching back 13.74 billion years to a singularity, we thought we had it figured out. Now, we seem to have found a major imbalance and have explained it with a cute new dark title, *dark energy*. Dark energy, like dark matter, is evident only in its effect in making space bigger.

Dark energy has no visible attributes and is the result of detecting faint light emitted by stars 10 or 12 billion years ago. We are now comparing light that was emitted a few thousand years ago to light emitted billions of years ago, (talk about snail mail), and concluding that the big bang was really just a mini fart compared to whatever is really blowing up our universal balloon. Hidden in all the formulae leading us to these almost preposterous conclusions is our misguided use of time as a term with significance beyond its actuality as a measure of comparative movements.

Relativity is everywhere and every-when and it is not a thing. Our standard for comparing rates and distances is the speed of light. We assume it travels at a constant speed through space even though we know it slows down traveling through transparent materials like water and in intense gravitational fields. The speed of light is our yardstick for measuring time and distances but what if our yardstick is elastic and stretches as the universe expands. We assume and have calculated a total mass for all matter in the Universe and assume and have evidence that the universe is expanding. Gravity also appears to be a constant directly associated with proximity and mass and the mass of all matter in the Universe appears to be constant and spreading apart creating a declining proximity. If the early Universe, the one sending us photon messages packaged billions of years ago, was less expanded, it would be more concentrated and photons, affected by a more intense universal gravity sum, would have started their journey at a slower pace, and accelerated as the universal gravity sum weakened. In the early Universe, our elastic light speed yardstick would have been shorter and would have gotten longer slowly as it made the ten billion year trip to our Hubble telescope.

We should also remember that what spit out the light long ago is long gone. We aren't seeing something that exists anymore, only a faint flicker of photons that have been traveling across space for a very long time and if they have not always been traveling at the speed we now observe, they could easily be leading us to false conclusions.

In another essay titled "The Inside out Universe" I propose an average energy density for space that may help explain gravity and if light is a universal variable it may help explain dark energy and if it varies throughout space it may also help explain dark matter.

33

PHILOSOPHY AIN'T DEAD YET

Philosophy gets a bad rap nowadays because it is misunderstood, not because it is useless and boring. Most of us spend at least a little time during a week philosophizing without realizing it. Don't believe me? Hang around a local bar just before closing, or listen to a mother trying to explain to a teenager why schoolwork is important. Philosophy gets a bad rap because we haven't updated its definition.

We still picture a philosopher as an old guy in white baggy pajamas setting on a marble porch with nothing to do but think up dumb arguments about dumb things. From our modern perspective the history of philosophy is bit like that, but philosophical thinking is still with us and very much alive. It just needs a new definition for us to recognize it. To begin let's make clear what philosophy is not.

Philosophy is not science. Science is a disciplined approach to examining, interpreting and describing the world and universe around us, including the "us" part.

Philosophy is not magic or sophistry. Magic and sophistry are slight of hand tricks meant to deceive the senses and lead individuals to false conclusions.

Philosophy is not Religion. Religion is an individual commitment to a belief in a supreme being or beings living on a mountain, or in another dimension, to which we are subservient and are obligated to worship.

Philosophy is as simple as looking out different windows in your house and describing what you see. Looking out an upstairs window presents a different view than looking out a basement window and we are all, looking out separate and different windows. Even when we look out the same window, we notice different things and describe what we see differently. It is interesting and historically informative to read ancient Greek accounts of what they saw from ancient windows and some of their viewpoints persist in our moral codes and forms of government

Philosophy is an art, the art of creating useful insights and perspectives. Circumstance, discussion, and use decide if an insight or perspective is useful. Formalizing the insight or perspective utilizes language as a paintbrush and, like a painting, it will be seen and understood differently by those who attempt to interpret it. A small point made just before the bar closes can hit home with an individual made receptive by a second or third martini and repeated in more sober terms, can spread and become a common perspective in a group of friends, a community or a country. Philosophy echoes through even simple exchanges of opinion or perspective and affects us all.

Like it or not, philosophy ain't dead yet.

34

Bond of Awareness

Empathy and cooperation among species become evident as common traits if we focus on the universal similarities of awareness instead of the great diversity in physical forms. Natural Selection has responded to the diverse and ever changing environment by allowing genetic adaptations to run free to produce microbes, magpies, minnows, mollusks and mice. It has been restrained in the creation of adaptive types of awareness by the limited ways in which living forms can sense and react to their surroundings.

Life can't exist, even in its simplest form, without some way to sense and respond to environmental situations and internal conditions. Early awareness may have been only simple chemical responses, (the gene being prompted to separate and replicate for example), but it has evolved in concert with biological complexity to become the eye, the ear, chemically sensitive buds and pressure and temperature sensitive cells. These essential sensory paths can be traced back to single celled life, some possessing primitive eyes some sensitivity to temperature some to vibration etc. From plants to mammals the natural selection of pathways for awareness, and the natural selection of nerve clusters to utilize environmental sensors most effectively, follow a common and narrow trace. We are related to all other living things more closely by our common ability to sense and respond than we are by genetic histories of form and function. Genetics has played a major role in the selection of our common abilities to sense, respond, learn, remember and choose. The limited ways in which we can gather and utilize information from our surroundings has made all life more alike in common pathways to awareness, than in any other.

Recognizing awareness in others, even other species, is the core of our cooperative and empathetic responses. Natural selection is not a formula or directive. Natural selection is simply the way in which life persists and adapts. The natural world around us has made us very different and very similar. To appreciate our similarities we need to begin to define our common awareness in basic terms. When we find life elsewhere in the Universe, classifying it in terms of its level of awareness may make more sense than looking for differences or similarities in form.

Emphasizing awareness over form may also allow us to dampen the inevitable reactions and religious revolutions the discovery of extraterrestrial life will certainly evoke. Humanity is the result of three billion years of naturally selected advances in awareness culminating in the ability to usurp natural selection itself and mandates that we think beyond ancient moral precepts. We need not give up our moral foundations, but as humans, (the only species determining the future of all life on the planet), we need to understand our responsibilities. Nature has empowered us with an advanced state of awareness and three mandates for its use: *Explore and learn*; *be of good council*; and *be a good steward*. The future of all life on our planet depends on our understanding the lessons of adaptation and cooperation taught by evolution.

35

HOW NATURE BUILDS THINGS

Combinational Directives

To build a concrete block wall we first have to build the concrete blocks. To build the concrete blocks we first have to make concrete and so on. There is an ordered series of combining steps required.

1. mix ingredients to make concrete,
2. make a mold and pour in the concrete,
3. let chemical reactions take place to harden the concrete,
4. separate the hardened block of concrete from the mold,
5. transport the block to the job site,
6. mix other ingredients to make mortar,
7. stack the blocks in an arranged order
8. bond them together to create a block wall

However, if the block wall is to act as a foundation for a house, we are not finished, and all subsequent steps also have prerequisite steps. You can't lay up the block before you remove the mold or before the concrete hardens etc. To reach a successful complex arrangement an ordered chain of combining patterns must be followed.

The proper proportions of cement and aggregates must be mixed and the mold must be of a standard size and shape to allow the combining process to continue. We observe the process of making concrete blocks and concrete block walls without amazement because we can see the entire process and have created the combinational directives that lead from raw ingredients to a block wall. This is what humans do.... they build things, but what about nature?

Nature is filled with combinational directives and, like our block wall, they are arranged in an ordered chain from the simple to the complex and, like our block wall, result in structures very different from the raw materials that began the process, but as we observe nature's methods, we should be amazed!

The combinational directives we are discovering in nature use enormous amounts of energy and raw materials, amounts far beyond our ability to measure or comprehend, and unlike our block wall, they are self-assembling. Nature's combinational directives are innate in energy and matter itself. The assembly directions are a part of the raw material itself. If we humans had the ability to infuse the raw ingredients of concrete with self-assembly instructions, we could stand back and watch concrete blocks form themselves and then assemble themselves into a block wall, (which would pretty much make us expendable).

Nature's combinational directives are truly amazing. They start with an expanding plasma that, (as it occupies a larger space), cools and precipitates basic bits of matter (fermions), then more energy participates condense becoming (hadrons) which create a new potential energy called gravity. Nature then continues its self-directed assembly by naturally creating composite elementary particles, (hadrons, baryons, and bosons) and further self assembles these into the first atom, (hydrogen).

With the basic ingredients self assembled, the expansion of the universe continues and residual energy is distributed throughout space as gravity, working in opposition to expansion, directs aggregates of hydrogen to gather into clouds. As gravity's grip grows stronger, the density of the hydrogen clouds increase and the cooling effect of the initial universal expansion is reversed as billions of small local hydrogen clouds become tiny pressure furnaces forging in their interiors, atomic composites from hydrogen. Step by self-assembling step, hydrogen is compressed becoming helium, helium into oxygen, oxygen into carbon and so on. At each step, the self-assembly releases some of the energy stored in the basic particles from their condensate origins, and with gravity holding these tiny furnaces closed, the process of atomic assembly continues until iron is produced. Iron acts as a stop production switch and shuts down the solar furnace. When Iron is fused, gravity wins the push pull contest it started in the first place.

Without fusion being able to continue, gravity collapses the star to become, either a weird dense ball of matter, a white dwarf, or an explosion, a super nova. White Dwarfs create more elements in their super dense cores, like silicon, and super nova explosions create even more, like gold and uranium.

One would think, that after all this self directed expanding, mixing and exploding, the process would be over, but nature's combinational self assembly directives have much more in store. Like our concrete block wall being only the beginning of a house, nature continues the construction process. Using a key player in Nature's combinational directives, gravity continues to gather clouds of hydrogen and scoops up the results of its compressed winnings from earlier exploded stars. Without restraint, it would seem that gravity would eventually win the collection wars by creating black holes and swallowing itself along with all matter and the game would be over, but gravity has another opponent beside the push back it creates by crushing matter inside stars. Gravity also has to contend with spin, the residual effect of being unable to gather misshaped cloud in straight lines to a central point.

Gravity's method of collection is messy. When gravity creates a star, it also creates rotation. Angular momentum as a natural byproduct and spins off the material it can't collect. When a balance is established between gravity's pull and the rate of rotation, the rotating material is held at a distance from the central gravitational point in a stable orbit, and orbits are everywhere. Our moon orbits the earth, the earth orbits the sun, the sun orbits around the center of the galaxy and galaxies orbit each other. Round and round everything goes. Orbiting appears to be, like gravity and star formation, another key combinational directive.

One would think that after all this additional mixing, exploding, gathering and spinning that the process would approach finality, but nature's combinational self-assembly directives have much more in store. Like our concrete block wall being only the beginning of a house, and the house being only the beginning of the activity within it, nature continues. The chain of events that has created atoms is filled with amazing steps, mostly violent, all taking place in very large numbers and over expanses of time so large we can't comprehend their durations. Nature creates using a self-actualizing chain of events directed by innate combinational directives far more complex than we can understand but can't deny, and they, like a congressional filibuster, go on and on and on

Beyond atoms, (has anyone ever tried to calculate how many atoms there are in the universe?) are the next combinational directed products.

- molecules, and from molecules,
- compounds and from compounds,
- minerals and from minerals,
- a really amazing self directed combinational product called life, and from life,
- awareness, and from awareness,
- a whole new chain of combinational directives as humans, (and probably other life forms in the universe), take charge of the assembly process.

Have all these amazing events existed as combinational directives since the beginning?

Was so much randomness requiered?

Were so many missteps necessary?

Is an end product programmed into Nature's combinational directives or are we, and other advanced life forms, the last step in nature's built in guidance program, leaving the future of the universe open and in our hands, or tentacles, or whatever.?

36

LEGISLATE, DON'T MANIPULATE.
A PLEA TO POLITICIANS

EXPLOITING DEMOCRACY

Politicians have always brought special agendas to congress for consideration. It is an essential part of any representative democracy, but it only works if the agendas are open for review by the voting public. Hiding priority agendas from the public during elections, cloaking them in misleading language and burying them in unrelated bills to become law, are subversive acts that undermine a free and open government. Some closed-door negotiations are necessary to coordinate political campaigns and to assemble blocks of votes, but underground campaigns to recruit and finance politicians in order to infiltrate our legislative bodies for narrow personal agendas is a subversion of democracy, not an exercise in democracy.

After many failures to gather public support for narrow views, extremists have begin to manipulate primary campaigns by sneaking their candidates into the system with carefully coached rhetoric, money from secret donors, and very sophisticated gerrymandering.

Secret agenda politicians concentrate on primary elections because primaries attract very little public scrutiny, have low voter turnouts, and are party oriented. With agendas hidden in party platforms, they are likely to carry forward to later general elections where puppet candidates are assured success, and take their assigned agendas to the next level.

The undermining process of rigging the system has resulted in a loss of public trust in the workings of what should be the most effective government system in the world. Strong views have a place in a democracy. Extreme views do not. Open debate and reasonable compromise are the bedrock of democracies but intolerance and strict ideologies is quicksand and will swallow all our freedoms. When pledges or religious beliefs, take precedence over an elected official's oath of office, the elected official becomes an appointed surrogate, not an elected official, and should be required to rescind their pledge or leave office.

"We The People" means all of us, not just the pious or the wealthy or the well educated. There is an innate wisdom in our diversity that needs to be respected by all lawmakers and put before fraternity, race, or religion. Stop preaching, stop blocking debates that might expose a weakness, stop pointing fingers and looking for ways to discredit others with different views, and stop fostering paranoia by demonizing others in order to get votes. If holding office, or your view on any specific subject, is more important than the integrity of our Country, you should not be in office. If you let competition supersede compromise in all debates, you should not be in office. If you love this Country, love all of it, not just your closed circle of like-minded friends.

Ask yourself why you are in office, how you got there, ask if you have the perspective and wisdom to govern well, and most important ask what or who determines your position on matters before you. Any answer, other than your own best judgment of what is best for the Country, should give you pause. We didn't elect you to represent Wall Street, your church, big business your political party or labor. We elected you to guide and govern a complex and diverse association of people with divergent ideas struggling in a complex world. You may not be able to understand all of us, but please try. Please put our best interests before your personal ideologies and political aspirations. Legislate, don't manipulate. Govern, don't dictate.

37

MOTION = TIME?

We have woven the idea that time is real into our formulas with misleading terms like *speed* and *velocity*, both of which are defined as distance divided by *time*. Formulae for motion have been so useful in identifying past and predicting future relative positions that we assume they also prove the legitimacy of the term; *time* but do they?

Having a three-legged horse listed on a betting slip doesn't prove the horse really exists. If we really want to be accurate, or at least complete, we should recognize "D", distance and, "T" time, are only comparative terms. The muzzle velocity of a bullet is described by our time/distance formula as 2,200 feet per second, and as long as we compare it to other velocities using the same time/distance conventions we create a useful and accurate comparative concept. For example:

A satellite in low earth orbit travels at about 26,400 feet per second or about 12 times faster than a bullet, but if we want a clear and complete concept of these velocities, we need to include the hidden comparisons used to arrive at the time and distance terms, (a second) and (a foot).

Most *time* increments are based on how far the Earth turns during an increment of a full revolution. For a second, it is 1/86,000th of a full rotation.

However, the rotating earth is a movement, not a time, and to arrive at a velocity, (feet per second), we divide a comparative length, (distance), *in this case the length of a king's foot*, by a comparative displacement, *in this case a small part of one full rotation of the earth*. To arrive at the speed of a bullet or a satellite, we divide an arbitrary length by an arbitrary movement.

Distance, like time, is only a comparison, and velocity, like time and distance, is only the derivative of other arbitrary comparisons. All we can do is compare. Everything is changing and nothing seems established, (*until we get to light*), and then what do we do?

We use the movement of a photon of light through space to measure both time and distance by inventing something we call a *light year*, and compare a photons travel distance through space to our planet making one trip around the sun, (which we learned above is composed of two other comparatives, *movement and distance.*

When we use the speed of light as a standard, we are comparing a comparison to a comparison to arrive at a standard measurement for a (*time/distance*) ratio, whatever that is? Even Einstein got tangled up with comparisons of comparisons when formulating his field equations and inserted lambda, a universal constant to stop the universe from becoming a variable, (a concept he later called his greatest mistake).

It is difficult to think outside the practical concepts and artificial comparatives that guide us through life. They have made us great engineers, but occasionally a human mind escapes from these conceptual traps and catches a glimpse of the magnificent swirling universe beyond our formulae and searches further.

We measure distances we cannot fathom, periods of time we

cannot imagine, and speeds beyond our comprehension, and delude ourselves into thinking we are close to a final answer. Even if we find a final answer, are we capable of appreciating it? Look at a star and try to appreciate the fact that you are observing, not the star, but a small trace of radiation emitted so long ago that the earth has rotated 73,000,000 times. Starlight expands in a sphere for observers everywhere in the universe, not just us. Have other observers also adopted *The Light Year* as a standard but based it on the orbital period of their planet? Are there any true constants in the universe, or only comparatives?

38

WHO STIRRED THE POT?

For a very long time we, (humankind), assumed that we were living on a static flat surface and the sun, the moon, and the stars were moving through the sky and then dived into the underworld only to reappear on the other side. For a very long time, we envisioned the sky as a dome with the sun dominant during the day and the moon and stars dominant at night. It was a grand show made especially for us and we accepted it in much the same way we later accepted the video game, *asteroid*. In the game, the little space ship disappears when it reaches the edge of the screen and reappears on the opposite side still moving at the same speed. I doubt if anyone ever turned the game over to see if the little space ship was racing across the back of the game to get to the other side.

As our ancestors continued to watch the sun and stars until curiosity finally got the better of them and, like my dog trying to squeeze behind the TV to see where Lassie went when she walked off screen, they took a closer look and started to track the motions above.

It took many stone monuments to get the intervals right and I wonder if we were trying to measure the heavenly movements or capture and control them by trapping them between bigger and bigger stones.

My dog still hasn't figured out where Lassie goes, but we humans have figured out why and how the sun, moon and stars move. It took a long time, but we have finally accepted the fact that we don't live on a flat fixed earth. In the early renaissance, explaining to the general-public, and to the church, how a round earth orbits the sun with the moon orbiting the earth was almost as difficult as my attempts to explain Lassie's TV disappearances to my dog.

Eventually, most of humanity got the new concept. Now, we throw ourselves into orbit, try to measure everything, and have found that everything in space seems to be going in circles.

I am getting older and occasionally I feel a bit dizzy, especially after a martini. I attribute these periods of dizziness to having spun around on the earth over 28,000 times since I was born and I suppose, being a bit dizzy is to be expected, but why all the circling and spinning. Planets spin on their axis and circle stars that are spinning and ride around in circles inside galaxies that are circling each other. I understand the concept that an ice skater turning slowly will spin faster when her arms are drawn in, but if she wasn't turning at all when she pulled in her arms nothing would happen. So what started all this rotating and orbiting?

Who or what stirred the Pot? Was the singularity spinning just before the Big-Bang? Did the singularity blow out on opposite sides like a fourth of July fireworks pinwheel? Or, was the initial angular momentum built up from the inside as a natural consequence of attractive forces as energy condensed into atoms, atoms were gathered into stellar dust by electrostatic attractions, and dust particles were gathered by gravity to form stars, planets, comets, asteroids and galaxies?

Even black holes spin. There is an enormous amount of angular momentum in the universe and as Einstein pointed out, angular momentum affects time the same way gravity affects time, it makes it run slower. It would seem that as gravity creates great centers of mass and causes them to spin ever faster it also slows all natural processes. Radioactive elements decay at a slower pace, I age slower, and as a fortunate consequence, I have time for one more martini.

But all this is occurring at macro scales. What about the spinning going on at micro scales? According to our atomic physicists, every thing is spinning in their world as well, and angular momentum is at the core of quantum mechanics where things don't need time to move about or change conditions. Whoa!

I think I will have that martini now.

39

WHAT'S HAPPENING?

"What's happening?" or "Wie Gehts?" or "的情况怎样?" is a familiar greeting used around the world. It isn't a sincere inquiry made to illicit a full explanation, if it were, we would be hard pressed to explain everything going on in the universe at that moment. The greeting asks only, what is occurring in your life that is significant now or in the recent past? It also assumes that the greeter knows your location. If I ask you "what's happening", and the last I knew you were in Denver, I expect your answer to reflect your recent activities in Denver. However, as we have learned in previous essays, position and time are comparatives affected by motion, and we have invented conventions that allow us to ignore most of reality's swirling confusion. If I ignore reality, and call Denver from London at 8:00 AM London time, I will wake you in the middle of your night. I am setting in the sunlight having a cup of tea, and if time were a real static thing you would be awake and having a cup of coffee, but instead you are trying to sleep. Your static position is relevant to my static position only in that we are sharing a ride on the surface of a giant ball that is spinning and traveling in an orbit around a star.

When I ask, "What's happening?" and you answer, "Nothing much." You make a universal understatement. We are bobbing up and down while we spin around in the outer arm of a spiral galaxy. We are also racing away from hundreds of billions of other galaxies, except Andromeda, (with which we will eventually collide) and, (because of our newly invented methods of en-masse communications), are evolving a new type of living awareness. If I accept "Nothing much" as a sufficient answer, I am not really expecting an answer at all, and if I explain this to you by pointing out that we are bouncing our voices off satellites in geo synchronous orbits, you will probably tell me to go to Hell and hang up.

The point is that we are tiny specs on a tiny planet lost somewhere in space busy feeding and breeding trying to remain oblivious to the discoveries of our curious scientists as they expose miracle after miracle for us to explain, or ignore. We don't have to climb to the top of a mountain on a clear night to marvel at the stars, or travel the world to marvel at the diversity and persistence of life because the curious among us have captured these marvels in language nets and made them ready, on demand, as digital information to be shared as synaptic realities. Our scientists, mathematicians and philosophers are fisherman, catching facts with intellectual nets drawn through inner and outer space by tools invented to let us see further and think faster. We have become a significant part of the many miracles that make up the magnificence around us and when another advanced awareness from elsewhere in space, eventually contacts us and asks, "Hey, what's happening?" How will we answer?

I hope that it will be with something more significant than, "Nothing much."

40

SIZE MATTERS

The complex interaction of movement and changing conditions in the universe is well beyond measurement or prediction. There is just too much going on. To compensate, we restrict our comparisons to things, situations, and positions that fit within our observational capability, (our scope of awareness).

As a way to isolate past relationships, and predict future relationships, we focus on small parts of the universe, and invent numerical relationships and formulae to compare changing positions in space or time. Two of these comparatives are distance and size.

A fixed distance is the space between objects moving at the same rate on parallel paths. *(If there is a true fixed point somewhere in the universe, we haven't found it.)* Everything is in motion. Even New York is in motion as it rotates with the earth at about 600 mph, goes around the sun at about 66,000 mph and rotates with the galaxy at about 600,000 mph.

A varying distance between objects occurs when objects are moving at dissimilar rates and/or are on dissimilar paths. This situation occurs so frequently at galactic scales we consider it the norm. We measure both fixed and changing distances using arbitrary standards created to make comparisons, and we have a lot of arbitrary standards, a foot, a meter, a furlong, a link, a chain, a mile, a kilometer, a parsec, etc.

To complicate things further we divide our standard comparatives into smaller pieces using arbitrary units like twelve inches to every foot and 5,280 feet to every mile, and for reasons we have long forgotten, we, in the US and England, have trapped ourselves into using these historical remnants. We have also trapped ourselves into using the convoluted mathematics that goes with them. More rational units, reflecting the girth of the earth, and are multiples of ten, are the norm in most of the rest of the world,

All the systems we use to compare things are arbitrary. There are no standard universal macro separation comparatives in nature. If universal standards exist, they exist at atomic or light speed scales and to be useful for our everyday applications become numbers with such very large positive or negative exponents they have little practical value.

Fixed distances, like the size of a room, (expressed in feet or meters), or the distance between cities, (expressed in miles or kilometers), are extremely useful comparisons. We are able to visualize these distances and we use them so frequently we forget they are arbitrary constructs

When we attempt to extend our common comparatives to very large distances, (like the distance to Mars or the nearest star), or to the very small, (like the size of a proton), we find our comparative conventions inadequate. Our concepts of distance, like our concepts regarding time, are useful, but if we want to understand more of the reality around us we need to remember that they are arbitrary constructs and avoid assigning them a reality they don't possess.

Distance has meaning only as a comparative that is relevant to the human scale of awareness or to the scale of a human inquiry. Time has meaning only as a comparative relevant to the human pace of awareness.

Are there any natural static distances or rates of motions in nature? Einstein gave up the idea of a universal constant calling it a mistake, but he included the speed of light as a constant in his formulation. We use the speed of light through space, 670 million miles per hour, as a standard in our cosmological examinations and consider it a constant, but is it? Has light always traveled at the rate we measure today or, when we examine light that began its journey billions of years ago, are we using comparative variable distances and movements as constants when, in fact, they have changed as the universe expanded? If so, we may be drawing false conclusions as to the actual age of the universe and as to its rate of expansion. Dark energy may be a false conclusion based on the false assumption that time has always been a constant.

41

WHEN TIME IS IT?

I find it curious that we do not have a proper interrogative to ask for the current time.

We use *Why* to get an explanation of purpose, *Where* to get a location, *How* to get an explanation of process, *Who* to get the identity of a person, *When* to get a time specified in the past or the future, and *What* to get an explanation for a process or an explanation of an object. We don't ask, "*When time is it?*" Instead, we ask "*What time is it?*"

It is as if we are asking for an explanation rather than a specific current moment, (and maybe we are). Time is a curious thing, especially since Einstein pointed out that acceleration changes the rate at which things happen and motion can go faster or slower depending on where you are. Even gravity can slow things down, but time is only a rate of motion and it is obvious that things don't all move at the same rate. Some things move faster and some things move slower.

The only motion limits we know are a temperature of absolute zero, where everything stops and the speed limit of light, which nothing can exceed. So what happens to our concept of time at these extremes? Curiously, time comes to a dead stop at both extremes and none of this makes any sense to a life form with a pace of awareness geared to the slow rotation of a small planet.

If we want to know what is happening on the surface of the sun we have to wait for the Earth to rotate 1/160th of a full rotation for the news to reach us. We have to wait is for the information to cover the 93 million miles between the sun and the earth. Light from the sun, and all other electromagnetic radiation, travels fast, but things in space are far apart and our only link to what is happening out there is from light traveling at 670 million miles an hour. Even at this speed, when we look at distant galaxies, we aren't looking at what is happening now. Instead, we are looking at what was happening millions or even billions of years ago. Current events aren't possible in space. Real time surveillance is not possible for distant objects.

To appreciate the distances and speeds involved an analogous model of our solar system may help. Blow op a weather balloon about the size of your living room and let it represent the sun. Then let a soft ball represent the earth and, to get it at its correct orbital distance, walk away from the balloon at a normal pace for nine minutes. The softball earth is now at approximately the correct distance from the weather balloon sun. If you want to extend this analogy to our nearest star, walk away from the weather balloon sun for two years. To get to our nearest galaxy, walk away for two and one half million years. The interesting part of this exercise is that, at the scale of our living room sized sun, you were walking at the speed of light. From this perspective, light speed seems like snail mail.

The most accurate clocks yet invented are two atomic clocks located in Bolder Colorado. They run at a rate determined by vibrating ions. The pace of vibration running these clocks, is also determined by their location in the earth's gravitational field. As an experiment, one clock was raised slightly above the other, which created a very slight difference in gravity and caused the higher clock to run measurably faster.

Additional gravity or higher G forces, alters comparative movements. Acceleration creates slower comparative movements in the accelerated object. If you want to outlive your grandchildren go on a long space voyage under constant acceleration, both on your way out and on your way back, or strap yourself into a centrifuge and endure a very long period of high G's, or move to the crushing gravity of a large planet.

You won't notice any difference in your heart beat or your aging process but they will be out of sync with those still living in a one G environment. To reverse the process, and let your grand children catch up, move back to Earth and put them in a zero G environment for an extended period. You need extremes to make any significant difference, but the effect is real and if you induce a difference it wouldn't make any sense for you to ask your grand children "What time is it?"

"When and where time is it?" would be more appropriate.

42

KILLING TIME

We measure time only as a comparative. We compare one interval of change to another.

If a ball takes X amount of time to roll down an incline and another ball takes less time when the incline is elevated, we make a comparison, a larger interval for the first ball to get from top to bottom and a smaller interval for the second ball. We observe the two-ball difference by using our human pace of awareness, which introduces a third comparative allowing quantification.

Ball number two seems to arrive at the bottom of the ramp in about half the time required for ball number one, but to be more definitive we need another more reliable and regular interval to use as a standard. Our most basic standard interval is the rotating earth and our latest, more accurate interval standard is the vibration rates of certain atoms.

The point is; the intervals we call time are only changes in position or condition. If nothing moves or changes its condition, including our synaptic awareness, there is no interval to measure and there is no time. Time requires motion and is meaningful only if compared to another interval of motion.

Fortunately, there is plenty of motion happening around us and comparative intervals are everywhere. The movements around us keep our awareness busy and delude us to into thinking we are passing through time when we are only observing, in synaptic intervals, motion and changes in conditions. Time is a construct created by awareness. Without awareness, even at its simplest level, even the awareness of a plant, time is only motion or a changing condition, nothing more.

Motion is displacement from one position to another but it, like time, needs another marked position as a comparative or it is meaningless. Motion without a reference point is the same as a static position. If you can't measure a displacement nothing moved and no motion interval was created. From this viewpoint time and motion seem so interdependent that they become the same thing. Motion, time and their derivative, velocity, are all one in the same and have meaning only as comparatives to other motions, intervals and velocities. Stop all movement and time also stops.

The earth rotates at a pace we call a day, and makes one revolution around the sun in an interval of motion we call a year. To be able to measure the blur of motion around us we have sliced up these standards to create smaller standard increments. We have chopped the rotational period of our planet into twenty-four pieces and then, to match other mathematical and geometrical constructs, have further divided these increments by sixty and then again by sixty to arrive at even smaller comparative increments of motion. We are now so accustomed to using these increments of earthly rotation to coordinate our activities that we loose sight of the fact that they are not real bits of time. The minute is a comparative we use to gage and coordinate our activities, but it is not an increment of time. It is the motion taking place around us while the Earth turns 1/1,440th of a full rotation.

We further delude ourselves by using our motion comparatives to specify when activities take place. We use words like *Now*, in the *Past*, or in the *Future*, intimating that these terms are descriptions of an entity called time through which we drag our awareness. We specify segments of motion comparisons by assigning numbers and names to the earth's rotational positions. For our 24 artificial rotational segments, we use numbers from 1 to 12 with AM and PM or 1 to 24 and consider all activities occurring during the specified rotational segment as an *hour* of coincidental movements. We further specify coincidental movements or activity by segmenting our 15 degree, *hour*, rotational segments into sixty smaller rotational segments of 1/4th of a degree of rotation, (minutes), and sixty even smaller segments of 1/240th of a degree of rotation (seconds)

We confuse comparative and coincidental motions with a concept (time) because we have filled our vocabulary with words connected to the concept, and because the concept is useful. Phrases like, *In the past*, *Two hours from now* and *Be there at four o'clock*, are a lot simpler and easier to deal with than, *Before we began this rotation around the sun*, or *When the earth has rotated 30 degrees*, or *Be there when the earth has rotated 60 degrees beyond the prime meridian.*

Our awareness has also led us to believe we are on a continuous passage from one moment to the next when we are really just interacting with the mix of movements we cause, those we observe, and those we remember. Time is a useful concept but October is only an arbitrary thirty degree orbital segment of the earth named after a roman emperor and a year is only one trip around the sun, and 4000 BCE is a religious marker indicating an orbit 6,016 orbits before the current orbit. Time is motion observed and measured against other movements, nothing more.

With time demoted to a construct of our awareness it no longer has any real qualities. Time becomes comparative motion instead of an invisible dimension through which things move. Time, as an original element of the universe, does not exist, only motion and energy fluctuations exist, and this mix of motion and change continues and is carried forward as a synaptic record, (if observed), or as mechanical interactions leaving traces of previous interactions, never stopping and never backing up.

Time as a construct of living awareness, began when chance motions and conditions produced a self-replicating, self-energizing combination of molecules called life. With life added to the evolving energy mix, the universe began keeping track of conditions and movements. Using replicating molecular combinations as monitors, bacteria began observing changes in the environment around them. Time is of significance only to a living awareness. Awareness, however, is of great significance and adds another dimension to reality. Self-awareness takes the universe even further.

43

BEYOND IDENTITY THEFT
- IDENTITY DEATH

Using a laptop as a weapon, skilled information manipulators are now capable of not only stealing your identity for the purpose of misusing your credit, obtaining a second mortgage, or filing a false IRS claim, they can also have you declared dead, raid your bank accounts and redirect life insurance payouts.

For the victims, this is more than an inconvenience. The victim is required to prove they did not initiate the declaration and claims, and prove they are still alive. Identity crimes are serious crimes. Crimes that demand serious investigation, victim compensation, and strong penalties for, not only the perpetrators, but for the institutions that allow themselves' to be manipulated and then refuse to offer serious help.

Identity crimes deserve the same media attention as violent crimes and the same investigative diligence. Identity crimes are no longer simple theft, they are being used as intimidation by opponents in court, by zealots to threaten individuals with different beliefs, and as a new intimidation technique to illicit protection money from individuals and businesses. Once targeted, a victim can be targeted repeatedly.

Changing one's SS # is allowed only in witness protection cases, and changing one's name is difficult and usually not effective. If you are pissed off at someone it is no longer necessary to kill them, instead, get them declared dead. Getting someone declared dead is more effective and involves much less risk.

Identity assassins are on the rise and an unsuspecting alliance has created a mafia type *pay for protection* between hackers and providers of protection software. (I sincerely hope this *hack to intimidate* and *pay to avoid* relationship is not intentional.

44

Why it Takes Two

Sex and death are intimately related. Before sex, life was immortal, or sort of. Before nature came up with gene swapping as a way to expedite the production of new attributes, life existed only as single cells that reproduced by dividing. One cell simply splits in two taking half the genetic material with it and both halves grow another half. Two new cells become nearly perfect replicas and the process continues. One cell became two, two became four etc, and no one had to die. Unfortunately, we humans can't do the divide thing and I'm not sure I would like to have myself as a roommate anyway.

Yes, twins, triplets etc. occur, and they too start at the single cell level. The difference is a separate step before the dividing begins. The disadvantage of immortal single cell division is slow adaptability. As the environment changes single celled life has to wait on a random faulty gene to produce, (by chance), an attribute that helps them adjust. Fortunately, single cells are good at playing the big numbers game and produce billions of cells very quickly by doubling repeatedly. If two cells become four in one minuet, it doesn't take long before ten thousand cells become twenty thousand cells in a minuet etc.

Gene swapping probably started when single cells with one attribute found it advantages to hang around with other groups of cells with other attributes and began to swap DNA as a way to make the union permanent, a kind of primitive group sex. These cell parties soon led to a continuing party and multi cellular life. Eventually, DNA transfer methods were perfected and multi cellular life developed complexity.

Even plants have sex, of a sort. Plants and animals need to live long enough to reproduce by gene sharing if their species is to persist. Immortality is still reserved for the single cellular. Gene swapping as a way of reproduction introduced both sex and death and sequential replacement has become a requirement for the multi cellular.

At the single cell level, one becomes two, two become four etc. At the multi cellular level, two becomes one.

It makes one wonder if sex is worth dying for.

45

AN ARMY OF ADDICTS

Armies enticed to fight by creating and controlling a substance addiction like heroine is a very disturbing concept. Addicts completely dependent on a regular supply of an addictive substance will do almost anything to get their next fix. They will steal from relatives, abandon their children and will even kill without remorse. An army of addicts would be fearless if they knew that engaging in combat or planting a bomb was a direct link to the drug that sustains them.

As far as I know, there are no examples of armies of substance-addicted combatants, unless you include ancient wine drinking groups like Joshua's band of killers. His small army of thugs slaughtered every living thing when they invaded a city and then spent a week of drinking and bathing to remove the blood and moderate their post traumatic stress.

Drug use is common in armies but not as a controlling element, or is it? Not all addictions are to substances like tobacco, alcohol, or cocaine. Some of our most powerful addictions are to internal endorphins, the chemicals that produce a sex drive, and sustain us as a race, but also drive some to act outside moral and legal norms by committing rape, sodomy, and murder. We are also predisposed to more subtle, but equally powerful, emotional tribal reactions essential to our early survival when we competed with powerful predators.

These tribal instincts still influence our behavior. Tribal instincts are a part of the histories of most of the world's religions including Christianity and Islam. Attend any worship service with an open and attentive mind and you can't help but observe these persistent emotions in worship ritual and pledges.

We continue to be embroiled in conflicts based on our instinctive tribal emotions as two of the largest and aggressive religions compete for followers and territory. This continual conflict began when Rome began to collapse and a new roman emperor became an advocate of Christianity, moved the center of roman power to Byzantium, and gave the young and relatively obscure religion status.

As an official state religion, Christianity replaced a long history of multiple Roman and Greek gods with monotheism. Shortly thereafter a displaced Arab tribal leader, who disagreed with temple practices in Mecca took a small band of followers to Medina and started a new religion, built a small army, returned to Mecca, and initiated an expansion of his beliefs that included strict rules and rituals. Those following these new proscribed beliefs quickly surrounded what was left of the Roman Empire with Islamic converts.

Adversarial tribal differences have been the root cause of many wars including the Crusades and now Isis. Christianity promises an after life to those who follow Jesus. Islam promises an after life to those pledged to Mohammad, and they kill each other inflamed by tribal emotions because they can no longer accommodate more than one god and cannot compromise.

Both religions cannot be correct. Only one monotheistic approach can be allowed, and our basic tribal emotions lead us to defend our beliefs from all foreign beliefs or ideas. We should be more rational in our modern civilized world, but both despotic regimes and democracies retain dogmas evolved from addictive tribal emotions sustained by religions. Monotheism is exclusive in its beliefs and adversarial in its practice.

A few more gods might make a big difference, especially if we didn't take them so seriously.

46

CREATION'S PATTERNS

When the creative patterns of energy and matter are seen as a preface to organic molecules, life becomes a natural step in the evolution of all things, and our advanced awareness becomes the natural result of an ordered process.

Life on earth could not begin before the earth was formed and the earth could not coalesce out of the dust of our forming sun before the long process of galactic formation had taken place. The sets of joining rules for nature's elements lead from the simple to the complex and from the inanimate to the animate. The rules also appear to lead from a purely mechanical universe to aware states. We are made of stardust and if we hope to find a place and purpose for our awareness, we must first decipher the process that has led to our awakening.

In the past, truth was revealed primarily in moments of inspiration or divine intervention. Our current perspectives are formed differently. They are the result of discoveries made using cooperative investigative efforts using sensory enhanced tools. With the help of our inventions, we are now able to look outward in space and back in time and find ourselves, not at the center of a small well-planned universe, but on a small planet orbiting a yellow star near the edge of an ordinary galaxy in an immense space populated with nearly a trillion other galaxies.

Using our sensory enhancing tools, we are able to look deep into the workings of the smallest elements of matter and find that we, and everything else in the universe, are composed of the same tiny particles. We have deciphered many of the joining rules for these particles and are beginning to understand how their joining creates emergent properties that lead to galaxies, stars, planets, life, and awareness.

Conceptual Patterns:

At first glance, nature's creative processes appear to be a disconnected series of events rather than a continuous process. We overlook the continuous flow of creation because our conceptual capabilities have programmed us to look at and understand our surroundings piece by piece and one event at a time. To resolve our intellectual disconnect with nature's holistic approach we search for small patterns that fit comfortably into our limited conceptual capabilities and then combine them to arrive at broader perspectives. We use analogy to reduce complex relationships to a conceptually acceptable image and add back complexity using symbolic tools like mathematics and logic. The patterns we use to understand the world around us may, or may not, be an accurate reflection of the reality we are attempting to understand, but nature used the same methods to produce our awareness that it used to produce our biological form. Our mental capabilities, like our bodies, have been naturally selected. Our conceptual abilities are limited but they compliment nature and allow us to play a conceptualization game "in-sync" with the way the universe is organized.

Before we began using scientific methods to examine the reality around us, our self-awareness produced only egocentric results. Early revelations and assumptions attempting to explain our place and purpose reflected self-aggrandizing perspectives. By learning to read the book of nature, our egocentricity has been partially set aside, and the comfort provided by human centered explanations has been replaced by an impersonal universe.

We now seem to be of little consequence and our advanced state of awareness has created a dilemma. Old perspectives provide comfort but seem inadequate. New perspectives seem relevant but empty of purpose, and we are left with a question.

Can we be both relevant and insignificant?

The accumulation of scientific information is accelerating and as our inquiries reveal nature's secrets, nature's methods are revealed. One of these methods is *natural selection*. Within nature's method of molding life and creating awareness, we may find a new grounding, a sense of purpose, and a way to integrate older perspectives into recent discoveries.

Atomic theory deals with particles 10 to the 25th power of ten smaller than us. Our cosmological explorations deal with sizes 10 to the 27th power of ten larger. We find ourselves miniscule in an unimaginably large universe and yet immense compared to its basic constituents. We find our short span of awareness to be a quick blink between the start of the universe, 13.7 billion years in the past, and an equally long period before its projected end, 14 billion years in the future. Our individual span of awareness is so short it seems meaningless.

Without the introduction of powerful analogies time extending in such extremes from our aware moment is well beyond our cognitive capabilities. We are here, between the largest and smallest of things at approximately the middle of time staring at the universe around us in amazement, exploring it in detail, questioning its purpose, and listening to the echo of our own self-awareness.

We are discovering that we are as much a part of the universe as the galaxies around us, and are composed of the same atomic particles. We have been created and shaped by forces and patterns we do not fully understand for a purpose we cannot discern, and which may not exist. Living things are the stuff of the universe evolving, an awakening dust gathered into complex forms struggling to understand their surroundings and themselves'.

The significance of stardust becoming aware is unknown, but if awareness is relevant it may give all living things purpose and human awareness responsibilities far greater than we have ever imagined.

Cosmological Patterns

Nature's processes flow through time as a complex mix of order within disorder, with energy becoming matter and matter creating energy. From observations, we have concluded that the natural tendency of both matter and energy is to become a diffuse smooth mix, a universe with a uniform temperature and no clumps, a diffuse atomic dust of equal composition, density and temperature. We have also concluded, from small scale experiments, that the total energy in the universe appears to remain constant, or is conserved, but this conservative smooth mix is not what we observe at large universal scales or at smaller atomic scales. The Universe is full of hot spots with basic elements accumulating into larger bits of matter at both micro and macro levels.

Our universal principals may eventually prove true at some extreme macro level as the universe expands, but when we use our observational tools to look back in time we find the uniformity that we expect to exist in the present existed only in the beginning. Instead of flowing from complexity to uniformity, the Universe seems to be progressing in the other direction. If our concept of entropy is correct there must exist one or more powerful forces at both the atomic and universal levels causing both matter and energy to concentrate into complex aggregations.

The patterns we observe at the largest of scales we group into a grand concept we call cosmology. These largest of patterns include the spontaneous emergence of all matter and energy from a singularity followed immediately by the mutual destruction of matter and anti matter and a diffuse cooling as the new universe expanded. We also deduce a non-uniform distribution of energy and matter due to some unknown initial influence and a shift from a dark opaque universe to one in which light escaped its dusty confines. These events were followed by the gravitational gathering of hydrogen to form immense sinuous clouds and the clumping of hydrogen clouds to form proto galaxies around central black holes.

The further gathering of matter within these spinning clouds formed the first giant stars, which in turn, created concentrations of gravity strong enough to compress and fuse hydrogen into helium and further into carbon and oxygen. Each crushing step producing enormous amounts of energy that held the enormous gravity of early stars at bay until the fusing process created iron. The fusing of iron from other elements produced no energy and allowed the giant stars to collapse. The implosion of such great massive objects and the additional fusion of elements by type 1A binary star systems, has fused even heavier elements and has distributed newly formed materials into space to be recollected forming smaller, longer-lived stars, planets, comets, moons, and asteroids.

This amazing series of events is still taking place in the changing event arena of time, a time that began at an infinitely slow rate of zero and is accelerating toward another extreme infinity. Time began at a zero rate (because of the immense gravity of the singularity), but time itself is accelerating as the universe expands and gravities universal grip is weakened. The time we use to measure events today is our current observable time, (measured by the speed of light at this stage of universal expansion). As the universe continues to expand, and universal gravity, (the sum of all material attractions), occupies a greater space, observable time will continue to be stretched toward an opposite infinity as matter becomes disbursed beyond interactive limits. The dark energy we postulate as the cause of the accelerating rate of *universal expansion* may be the acceleration of time itself as the universe slowly evaporates.

Atomic and Molecular Patterns

To understand the great variety of matter we find within observational limits we group the patterns into particle physics and chemistry. We have identified the basic building blocks of matter making up atoms, and have uncovered a hierarchy defined by ordered combinations of smaller components. We have also created a visualization tool to understand this hierarchy by placing atoms into categories based upon atomic number and valance. We also understand atomic joining rules and the basic structure of the materials their joining creates. From these inorganic joining rules we have extended the patterns to a grand concept for organic molecules and identified the carbon based congregates that make up life and have traced their natural joining tendencies.

Complexity and diversity increase exponentially at every level. Quarks combine in a limited number of ways to produce protons and neutrons. Electrons combine with protons and neutrons with more latitude to produce a hundred or so *natural* atoms. Atoms combine with each other in many more ways to produce an extreme variety of chemical elements that in turn combine to make up the observable and extreme variety of organic and inorganic matter.

Organic Patterns of development

To understand living forms we group the patterns we observe in nature into a grand concept called biology. To accommodate our conceptual limitations we divide the totality of living forms into kingdom, phyla and species. We use these categories to trace life's origins and its evolution and to decipher the chemical codes that direct and control the process. Developmental patterns include a myriad of interdependent relationships.

- environmental relationships
- group adaptability
- symbiotic relationships
- parasitic relationships
- predatory effects
- competitive influences
- cooperative influences
- chance

Life sciences now dominate much of our scientific efforts and reflect our continuing need to answer the universal questions imposed by our self-awareness.

Patterns of Awareness

Still missing from our basic set of explanatory categories are explanations for the combinational and developmental patterns of awareness itself. Initial efforts to identify the genes responsible for organs of awareness and to trace their origins are underway. Links between genetics and behavior are being studied, but questions as to how awareness springs from a physical form, or how the complexities of awareness have evolved, and how the evolution of awareness relates to genetic selections, remain unanswered. We have not grouped patterns that help us understand awareness into a grand concept because scientific methods are best suited to the examination of nature's physical aspects. Progress is also difficult because, attempts to create hypothetical explanations for the development of awareness are easily misdirected. Awareness, as a concept, may be beyond our current *inductive* scientific methods but, If we are careful not to stray from a serious examination into an egocentric analysis, it is fair game for deductive examination and philosophical inquiry.

47

A Question of Perception

We live together as couples, families, co-workers, tribes, congregations, and communities. We also live together in cities, counties, states, countries and on the same planet. Humanity, when it is stable, is a great cooperative endeavor. A productive effort that creates civilizations, technology, advances in living standards, better health, and security. When humanity is unstable couples argue, politicians slander each other, governments disagree, and wars occur.

Disagreements occur over scarce resources or disputed boundaries but more often, disagreements occur because of a difference in perspective. Be it personal, political or religious, disagreements in perspectives are more difficult to resolve and can result in physical confrontations lasting for centuries, especially if the differences are ideological. Differences in perspectives can also fester and become killing contests bent on reconciliation through elimination or submission.

Binding us together in cooperative alliances arrived at by reason or war, are commitments that take several forms and have many names. These can take the form of a treaty, an oath or pledge, a religious directive, or an imposed mandate.

To be lasting, our cooperative allegiances need to be flexible and adapt to changing conditions. Unfortunately they are, too often perceived as immutable, and carry within them the seeds of the next confrontation. An unwillingness to compromise on an oath or pledge, or an unwillingness to alter one's perspective regarding whose god or revealed commandments should prevail, or an unreasonable attachment to an outgrown perspective, or poorly defined mandates, continues to slow humanities progress and provide the excuses needed to continue killing each other simply to prove a point.

Democracy is an experiment begun thousands of years ago, and although it reflected only the free choice and opinion of ten percent of the population, it was a step away from the absolute authority of a single individual. We, in the US, are continuing the great experiment with a majority of the population participating and the voices of most citizens heard.

We remain a Country divided by income and beliefs, but ameliorate our differences through the rule of law. We hope to remain disciplined and sort through our differing opinions by trusting our elected officials to put their oath of office above any pledges or narrow ideologies. Unfortunately, we have begun to question if our trust is justified.

Our democracy has been put at risk by those who run their campaigns, and exercise power using paranoia, slander, innuendo, and misinformation. We go to the polls, not informed on the issues and the real abilities and opinions of the candidates, but stirred by angry rhetoric to vote against individuals out of hate, not for the best qualified based on an informed opinion.

We are herded, like well counted sheep, to vote in primary elections with the polling places carefully gerrymandered to insure outcomes, and are discouraged from voting by statistical forecasts. Democracy is a matter of perception, not as perceived by a few but the combined perceptions a diverse society participating and being represented.

Democracy requires a great collaborative effort to define itself, its goals, and its relationship to other nations. It is not a sporting contest between teams with different colors and different mascots scoring points on every issue as if striving for an ultimate victory.

Our future is not a choice between the *R's* the *D's*. It is a choice between confrontation and collaboration. We all win or we all loose, not by defeating those with differing perspectives but by cooperating.

48

LINES AND CIRCLES, SCIENCE AND RELIGION

We live in a world defined by lines and circles. Lines on the ground define Countries and their borders. Lines also define divisions within larger borders as territories, counties, districts, cities etc. Other lines define property ownership, easements, and use restrictions.

Within each of these many boundaries, different laws and rules apply and disputes over lines are frequent, especially when the rules themselves' are in conflict. In spite of the confusion, most entities defined by lines find ways to cooperate and trade, but tension is always present.

In addition to divisions created by lines there are divisions created by circles. Circles, not drawn on the ground, but in the mind. Organizations of like-minded individuals persist in as many diverse ways as those defined by lines on the ground, but have different types of boundaries. Religions reach across *line boundaries,* and are divided into many smaller circles. Fraternal organizations, associations, brotherhoods, and many others exist and they too are often at odds with each other and themselves. Mutually pledged groups remain committed to one another, not by laws or rules of government, but by a common belief, a goal, or an enemy. Each focusing inward on narrowly defined perspectives.

Blurring all of these lines and circles is the exponential growth of information, much of it new information resulting from our careful examination of Nature. The scientific method of theorize, verify, and confirm, has resulted in new technologies that flow across borders, invade circles of belief and create new common perspectives strengthening trade arrangements but undermining belief based perspectives. Science and technology threaten narrow perspectives with a wider view.

We are engaged in ideological conflicts between many divergent belief systems and between belief systems and secular perspectives. Nearly all belief-based groups, large and small, define their tenants and rituals carefully in order to maintain continuity, to differentiate themselves from other faith-based organizations, and to separate themselves from those with secular perspectives.

Closed circles of belief depend on narrow governance and strict obedience; are well defined and clearly express their differences. The religions fathered by Abraham have splintered into Islam, Judaism, and Christianity, and these have further splintered into several Jewish faiths, Greek Orthodox faiths, Roman Catholicism, Thousands of protestant sects, and Shea and Sunni beliefs. All of them are confrontational in some way or another. Some have become so narrowly self defined that they are at war with everyone.

Faith based wars are as old as civilization and none are blameless. The believers in Moses as the great lawgiver are now at odds with those who believe just as fervently in Mohammad as the great prophet. Moses led an armed group of rebels out of Egypt and in following generations razed city states in brutal attacks to make way for the twelve tribes of Israel. Many of these attacks were even more vicious than the recent atrocities of ISIS. The Christian conquests in the holy lands, the Persian conquests, Mohammad's militant return to Mecca, and many others, all attest to faith based conflicts and they are not over.

We now live in a technological age. An age based on scientific investigations that have fostered tools that allow us to enjoy luxuries and to communicate in ways unimaginable just a few years ago. Our scientific tools allow us to look deep into space and into the constituents of matter and have resulted in a new and more realistic view of our place and purpose in the universe. Our expanded perspectives are well beyond those of Moses or Mohammad, but faith based organizations still cling to simpler explanations and restrict access to wider views, create counter arguments for scientific explanations, and view new perspectives as threats. Unfortunately, organizations shielding themselves from expanding perspectives may also be hiding from implied new social responsibilities.

We are now aware that we live on a very small fragile planet circling an ordinary star near the edge of a galaxy along with more than two hundred billion other stars in a universe with a trillion other galaxies. Those unwilling to look,

- at the sun as a roiling ball of hydrogen with a nuclear furnace at its center,
- or to look deep into the darkness of space,
- or at the simple way four amino acids spell out an adaptable system for life to adapt,
- or at themselves as the product of an amazing chain of trials and errors,

miss the true miracles around them and discount the challenges and responsibilities that our advanced state of awareness imposes.

Faith based organizations insist that moral behavior can be maintained only by the threat of punishment, or the promise of a reward after one dies. In a secular universe, morale behavior exists in nature and in humans without the need for a dictatorial belief system. Empathy and the impulse to cooperate and communicate have been implanted by evolution and when we peek out from under the security blankets of religious promises and begin to appreciate the wonder and immensity of the reality around us, we begin to recognize the greater opportunities and responsibilities our evolved position offers and demands.

Beyond the behavioral directives of gods, nature has given us three mandates:

- (Explore and learn), our intellectual abilities have taken 3.5 billion years to emerge from the dust and are a fragile spark to be used wisely
- (Be of good council), Language in its many forms is our primary tool for both discovery and cooperation and when subverted for selfish purposes seeds self-destruction.
- (Be a good steward), We have usurped natural selection with our advanced state of awareness and are now responsible for our own future, the future of all life on earth and the planet itself. We are stardust awakened and mandated to begin the next phase of evolution directed not by genetic chance, but by intelligent design, *OURS*.

49

You Can't Disprove What Can't Be Proven

There is a continuing dispute over what constitutes proof. The disagreement is often between science and religion.

Religious proof for a deity, life after death, or the relevance of prayer comes from revelations or directions given by a deity to a chosen-few, and is substantiated by a chain of supporting testimony.

Proof for science comes from observation followed by theory. Proof for a theory requires substantiation by further observations, experiment, mathematics, and continuing critical reviews.

Both systems of proof create changes in the doctrines they support but in different ways.

Religious doctrines change as social structures evolve and doctrinal leaders reinterpret and add or remove dictates, rituals or portions of holy writings.

Scientific doctrines change as new discoveries disprove or replace existing theories.

One changes by dictate, the other by discovery. One is relatively static, the other fluid. One claims responsibility for creating social order, the other for creating the technology that supports modern civilizations.

One claims an exclusive access to truth. The other makes no such claim and expects its findings and theories to be modified and replaced as new discoveries are made.

Disputes between science and religion are common. A resolution of conflicts between religious views and scientific findings is unlikely and arguments futile. Those who depend on faith for final answers base their arguments on irrefutable information that requires no demonstrable proof. Those that accept scientific evidence understand that knowledge is constantly growing and adjusting as discoveries expand humanities knowledge base.

The uncompromising absolute right hand of religion and the always-compromising left hand of science occasionally find commonality through compartmentalizing, but even in these rare agreements, the rigid right is rarely able to compromise completely with conflicting scientific findings. Religion depends on reveled truths that are beyond the demonstrable proofs required by science, and science refuses to accept what cannot be verified. Any argument against a faith-based tenant is futile because you cannot disprove what can't be proven.

If it were possible to disprove life after death, would suicide bombers think twice? Would church donations stop? Would people stop praying and become more responsible? Would the United States Congress suddenly become rational?

50

PPD Politics

Politics is played differently in different forms of government. The pursuit and control of power has evolved from hand-to-hand combat by rival tribal leaders, to long successions of kings, to a pandering to the multitudes. As the paths to power have evolved, so has politics itself.

Democracies have existed in many forms and those existing today each have their own unique organization and style. Some are shams created to cloak a despotic regime, some are restrictive allowing only partial participation, and some are all inclusive allowing the opinions of all citizens to be considered,

Our special form of democracy is but one of many. It has a special history and a constitution based upon a reasoned attempt to distribute and restrict power to prevent a despotic take over, and to protect its citizens from unreasonable oversight and control. It has adapted to changing world conditions and the impact of new technologies. The government established by the US constitution has served us well in spite of having been tested often by individuals attempting to interject narrow and restrictive ideologies into the democratic process. When this occurs the representative process is compromised.

Ideologues trying to control our democracy use statistics to manipulate voting in primary elections, use "K" street operatives to predetermine the positions of elected officials, and use misdirection and misrepresentation to sidetrack legislation. Driven by their narrow views they hide behind religion to misdirect any reasoned argument and rely on emotional appeals using PPD politics, *(Prejudice, Paranoia, and Denial)*, to justify their actions and to create a following. Hate is easily aroused as is paranoia. Both are powerful manipulative tools when facts are not advantageous and denial is an effective alternative, if the facts are difficult to refute.

PPD politics have a long and dangerous history, have led to many atrocities, many wars and is now spreading across the globe under the guise of resistance to a radicalized religion. Democracies are based on reason, (not religion), law, (not leverage), and inclusion, (not exclusion).

Hate, Fear and Denial are spread easily and are always dangerous. Those who rely on these emotions as tools in the pursuit of power have no place in a government based on the ideals that all men are created equal and have a say in how they are governed. PPD politics, (prejudice, paranoia, and denial) destroy democracies from the inside.

51

A MESSAGE FOR THE PRESS AND POLITICIANS

I like to think that our elected leaders and the free press are there to find and correct, or at least suggest, corrective methods to problems we encounter as a free society. Over the past few years however, I have observed a disturbing trend of ferreting out oversights and mistakes, not to correct them, but to exacerbate situations for selfish reasons.

Two self-aggrandizing examples stand out.

1. Reporters and editors exaggerating situations to gain ratings,
2. Politicians; vilifying individuals and opposing office holders to gain political advantage without a thought to the divisiveness and paranoia they are creating.

Some even advocate a sort of revolution against the framework of our democracy, and have signed pledges to reduce the size of our government to something they can drown in a bathtub and I have been confronted often by individuals who have been seduced by these purveyors of hate.

It is acceptable to have differing opinions and to be angered by dishonesty and deceit, but using every problem or misstep to gain personal advantage, or as a way to promote paranoia and distrust, is the lowest form of patriotism and boarders on treason.

If we have a problem, help solve it. If someone makes a mistake, point it out and move on. Our democracy has to be rebuilt every day as our civilization becomes more complex, faces new challenges, and new enemies. If you can't respect the person holding office, at least respect the office. A self-aggrandizing press, self-serving politicians, and a misled populace will never realize their dream if hate prevails, democracy doesn't work that way. When you find a problem report it, focus on it, and fix it without grandstanding, only then will we watch your news and give you our vote. We aren't as dumb as you think we are.

52

NOW AND THEN

In casual conversation, we use the "now and then" reference to mean occasionally. Ask someone how often they trim their toenails and their response will often be; "now and then". But the "now" and the "then" if given as an explicit answer, thinking the question was more than rhetorical, could mean, I am trimming them now and I trimmed them at sometime in the past. We don't make this mistake because we understand that the question is not expected to illicit an exact answer, (like every 17 days), and we expect the answer to be non specific.

When we get to questions asked of nature through observation and experiment however, the answers are specific, not allegorical, and are often difficult to understand. Nature's answers are complex and when they exceed our conceptual limits we return to an acceptable conceptual allegory. We simplify in order to understand.

A good example of this is our current cosmological assumptions of the age and size of the Universe. Until we can identify and measure gravitational waves, we have only one energy source available that we can use to interpret the reality beyond our atmosphere, *(light)*.

Fortunately, most of the objects in space either radiate or reflect electro magnetic radiation and to further simplify our observations we discovered that all types of radiation travel at the same speed through space and are made up of measurable frequencies and intensities.

Using these discoveries, we have identified various atomic elements from the spectrum of light they emit. We have also measured the approach or departure speeds of emitting objects by the altered wavelength of the light they emit. From this one vantage point, using only one complex signal source, *(light)*, we have constructed an entire image of the universe around us, its size, its age, and have postulated its beginning and forecast its eventual demise.

Using our optical, radio, infrared, x-ray and microwave, telescopes we have looked out as far as we can, and back in time as far as we can, or have we?

We can't ask the universe when she trims her toenails because, as far as we know, she doesn't have any, but we have asked her how old she is and she has answered modestly,

"As old as the dimmest light you can find".

Her answer is much like the answer we get from humans when we ask about their toenails, "Now and Then". The "Now" part of nature's answer, as to her age, is the light we can see now. The then part of nature's answer is the light we can't see, light that hasn't reached us yet.

For us there is only an observational "Now". Light that, according to our assumptions about its speed through space, may have started thirteen billion years ago. We aren't looking back in time, per-see. We are only looking at light that started its journey long ago and has collided with our observational "Now" at the time of our observation. For us, inside the universe, asking questions, there is no "Now and Then", There is only "Now" and an assumption.

53

THE AGES OF LANGUAGE

We categorize most intellectual information to facilitate selective studies and learning efforts. We divide scientific endeavors into the physical sciences and life sciences, each with many sub categories, and these divisions remain fairly static. The categories in the study of human history however, change whenever historians impose their own perspectives.

Will and Ariel Durant divided history into:

Our Oriental Heritage
The Life of Greece
Caesar and Christ
The Age of Faith
The Renaissance
The Reformation
The Age of Reason Begins
The Age of Lois the XIV

Other categories abound and each reflects the unique perspective of the scribe recording the activities of the past. I am no exception. My preference for categorization is to divide human history into periods differentiated by the preferential uses of language, not which language was preferred, but how language was preferentially used to express the most common attitudes and perspectives of the time. My categories are:

The Age of Self Aggrandizing Languages
The Age of Commiseration
The Age of Allegories
The Age of Directive Languages
The Age of Expanded Meanings
The Age of Expanded Communication

In an essay of reasonable length, and considering that I do not have enough years left to duplicate the investigative detail of Will and Ariel Durant, I offer only an overview of why, and how, my categories are indicative of a separate age, and hope others will follow my thinking to see if the categories have merit.

The Age of Self-Aggrandizement is long and encompasses the early development of language to include titles and self-aggrandizing terms used to promote the self importance of those holding social power. From early tribal leaders to Pharaohs the languages of this period are rife with honorific and aggrandizing terms reflecting the perspective preferences of the period for inclusive social organizations under a strong leadership.

The Age of Commiseration began with the introduction of monotheism and expanded when the follow-on idea that a divinely appointed human might not be needed for access to god. Groups separated from long established aggrandized empires became outcasts and as they struggled to establish their identity, their language evolved to include many self-deprecating words and phrases. The Old Testament is as much a testament to the struggle of the dispossessed as it is to the adjustment of their language to express their attitude of being the suppressed chosen few.

The Age of Allegories began as Roman rule expanded into areas where languages had become truly self indulgent. As Latin began to erode the languages and perspectives of peoples not accustomed to the harsh factual meanings of Latin, those still using native tongues began to use allegories to evangelize and to avoid repercussions from Roman rulers. Being less direct, a message or lesson given as an allegory seemed more like an example of a hidden truth than a factual comment and quickly became preferred method of speaking and writing. The New Testament is evidence of the increased use of allegories during this period with both Jesus and Paul using allegories extensively. Jesus used them to advantage by turning any event or situation into an allegorical message. Near a well he would speak of the water of life. Near the sea he would speak of fisher's of men. During a meal he would talk of symbolic eating and drinking. Jesus spoke only to the Jews but as Paul took Jesus' message further he followed Jesus' example and filled his letters to new churches with allegorical examples and lessons.

The Age of Directive Languages began as Europe developed a multitude of kingdoms and as tribal disputes in The Middle East were resolved to become the Koran. The age of directive languages continued when the Bible was translated and transformed by secular rulers anxious to take advantage of the appeal of religious messages promising an afterlife and promoting compliant behavior. Directive passages are prevalent in the Koran and are emphasized in revisions of both the Old and New Testaments.

The Age of Expanded Meanings was born of necessity as naturalists and scientists began to expose a reality we had long ignored while caught up in religious perspectives directing a head down eyes closed perspective. When a few brave men disobeyed church directives and began to ask nature questions using the scientific method of test and verify, nature willingly gave up her secrets and with mountains of new information to deal with, languages had to expand. A dictionary from the twelfth century is much smaller than a dictionary from the twentieth century.

The Age of Communication has grown out of The Age of Expanded Meanings. Discovery and exploration yield technological advances as a natural consequence of exploration using new tools, and the spin-off of new discoveries using new tools is new inventions for communication. From this cooperative creative system we now can talk to anyone on the planet using a small device we carry in our pocket, are able to stay informed, to learn and be entertained by thousands of communicative channels. Human behavior is changing in ways we cannot predict and language is changing to accommodate our new perspectives and social arrangements.

54

ALTERNATE EXPLANATIONS FOR DARK STUFF

I understand why Einstein inserted an extra term into his general relativity equation to keep the universe static. He simply didn't like the idea that the whole universe could be a variable. If he were alive today I doubt he would like the idea of dark matter and dark energy any better, and I have to agree. We are basing a lot of our latest theories on assumptions we have come to accept as facts, more specifically, "the speed of light" and "mass/gravity equivalence"

Not being a mathematician, physicist or cosmologist, I can think outside the box with impunity and, without peer reviews to worry about, and having not been indoctrinated, I can wander around outside the box and look for new perspectives, (something I'm good at after three marriages and five careers).

Starting with dark energy, (*ironically discovered using a space telescope named after the astronomer that first identified galaxies as objects outside the Milky Way, all receding from each other*), we now see the galaxies furthest from the creative moment moving away faster than those closer to the creative moment. (The older the galaxy the faster it is receding from us and other galaxies). Our conclusion is that the rate of expansion of the universe is accelerating. We use the red shift of light to measure our observations and confirm them by measuring luminosity from Cepheid variable stars and type 1A novas. In short we are using light as our standard of measurement and are confirming our results by measuring light a second time.

Light is all we have. At the center of all this measuring and confirming is an assumption, *the speed of light is constant in a vacuum and has always been traveling at 670 million miles per hour.*

But what if light is a variable? On one of my trips around the outside of the acceptable scientific theory box, I noticed that one side of the box seemed shorter than the others. All of the sides had a light yardstick taped to them that I could detach and use. I used each yardstick to measure its corresponding side and found all sides the same, but when I used a yardstick from a different side I got different results. The light yardsticks looked alike but when I compared them, they were of different lengths. I also discovered gravitational anomalies that were affecting the length of the yard sticks and thanked Einstein for making the relationship between light speed and gravity clear.

I tried explaining what I had found to the scientists inside the box but they didn't want to accept the idea of a universal gravitational coefficient that decreased as the universe expanded making light yardsticks from an older and more condensed universe shorter that the light yardsticks we use now. If the speed of light is a variable, our measurements both of red shifts and luminosities are skewed and are deluding us into thinking the universe is expanding at an accelerated rate, making dark energy an unnecessary construct.

Getting nowhere with an alternate explanation for dark energy I resumed my outside the box explorations and encountered two unusual spinning black tops, (like those children play with). Both tops were in a vacuum, one was tiny had a weak gravitational pull and a slow spin rate. The other was large had a huge gravitational pull and a higher spin rate. As I watched the tiny black top a gas cloud drifted by and began to be drawn in toward the toy. As the gas gathered tighter around the top the gravity at the center increased and the spin rate at the center increased faster than the spin rate of condensing particles further out until the gas was all gathered and the spin rate stabilized into an orbital curve typical of most solar systems with distant objects orbiting slower than objects orbiting closer in.

Then I turned my attention to the large black top as it flipped on its side, as tops do when their spin rate slows, and began to spew gas out of both its polar axis creating a pinwheel of gas and condensing materials rotating in concert with the top.

Both tops were now at the center of miniature galaxies. One of the mini galaxies grew from the outside in. The other grew from the inside out. One rotated as a planetary system, but the other mimicked what we observe in the symmetrical rotation of spiral galaxies. Could it be that the mysterious super massive black holes we are finding at the center of nearly all galaxies did much more than form from in-falling gas and debris, and instead, early in the creative evolution of the universe, were exploding pieces of the singularity, secondary sparks of creation, like the last dazzling spark display at the end of a fireworks show. I'm not even going to try to present this idea to those inside the box, but it is one way to explain away dark matter. Inter galactic dark matter has another explanation but I've been told by several scientists to stop walking around the box of secured ideas.

As a tribute to Einstein, I am calling my unformed theories, Lambda 2 and Lambda 3

55

CAN ONE THINK AND EMOTE AT THE SAME TIME?

In my youth I was instructed by a teacher, "You cannot think and emote at the same time".

Until recently, I gave that instruction little thought. But prompted by recent terrorist acts, inflammatory political speeches, and changing social attitudes, I have reviewed my own past and my own changing attitudes. My teacher's, separation of emotion and reasoned thought as coincidentally incompatible, have many examples in my own life. Their incompatibility has restricted the full expression of my potential and I am ashamed of most of my misplaced emotional responses. History too, has many examples of emotions overriding reason and limiting the full potential of the human race.

Passion has many faces, including religious fervor, physical attractions, hate, and extreme dedications to narrow causes. Being caught up in one or more of these chemically induced states is not only seductive and directive, it also overrides a large portion of one's thought processes. Being filled with religious fervor makes one receptive to directives offered by orators skilled at arousing compelling emotions.

It is not 'thought' that is displaced by emotion, but reason. Emotions narrow one's perspectives, pushing reasonable evaluations made using wider views aside. Being convinced emotionally that non-Christians were sub human, allowed the Crusaders to slaughter thousands with differing perspectives without compunction. The Crusaders had their focus narrowed by emotionally intense instructions and rituals and were freed from a sense of guilt and disgust for their actions by having their actions absolved, in advance, through ingrained emotionally misdirected responses, numbing them to the carnage they caused.

Humans have several susceptible sensitive areas where emotions, when evoked, can override reason. Sex is the most obvious, followed closely by tribal and survival responses. In early humans, reason played a lesser role in both individual responses and societal arrangements. Impulse, instinct and emotions acted as the primary guides for human survival. As societal arrangements became more complex more protective and more productive, reason became more important and began covering innate emotional survival instincts with a thin veneer of constructed societal norms.

All of us have these thin applied areas of self control covering our more basic response mechanisms. Over time, these early layers of new social norms became evolved attributes that have brought us successfully through our period of conversion from animal instincts to civilized states. The societal veneer that directs our behavior today goes beyond these early imposed controls and is learned. It is implanted after birth and must be instilled in every individual if that individual is to successfully integrate into society. We are born with only a few proper tribal/family behavior patterns and compassionate instinctive responses. The rest must be taught by example and instruction.

When we fail to instill this veneer of societal integration and fail to reinforce it with logic and expanded world views strong enough to resist emotional calls for misdirected behavior, we create displaced individuals that become susceptible to despots.

A sense of history, a basic understanding of geography, a basic understanding of government and societal organizations, an appreciation of language, a basic view of comparative religions, a basic understanding of economics, and of science and technology are necessary prerequisites for reason to resist subversive appeals to our primitive instincts and emotions.

Unfortunately, ideologies, polite exclusions, and religious disclaimers have purged curriculums in private, parochial and public schools of many of these basics and made less informed generations vulnerable to the seductive call of narcissists and megalomaniacs. By dissolving the thin layer of protective logic with caustic rhetoric, despots are able to tap into our primitive emotions and alter our individual prime directives. "Fear your government", "Hate the Jews", "All Muslims are terrorists", "Only Christians are moral and can be trusted", and many other appeals are made chipping away at the thin veneer of logic and understanding that sustains our modern global civilization.

Is it possible to think and emote at the same time? It depends! It is a matter of choosing between, ancient beliefs and discovery and education, between emotion and reason, and between confrontation and compromise. Do we want to return to the murderous 15th century's religious conflicts, or struggle to support a modern civilization? Compassion and love are emotions we value, but they will only survive if we are not afraid to face truths beyond dogma and allow reason to prevail. We can give your children a gun and teach them to defend the beliefs we instill, or we can give them a sense of worth and a sense of belonging in a much larger universe, and teach them to reason.

56

Rare and Responsible

The Universe exists in two places at once. It exists as matter and energy interacting, (a universe of mechanical probabilities), and as an interpreted collection of observations by aware organic forms. The interaction between these two states of reality, (and both are real), is one of dependence and modification. Organic forms are created and are dependent on limited interactive probabilities produced by random atomic and celestial arrangements. Organic forms remain dependent on these random arrangements and are subject to modification and elimination as the tenuous balance of matter and energy that allow them to exist morph or is altered by a new interaction

To a lesser degree, the universe of mechanical probabilities between matter and energy is subject to modification by aware organic forms and, on small scales, can have its random possibilities directed.

Organic aware forms have only limited power, using collections of observations as tools to alter mechanical states but, at advanced levels, can alter the local random balances of energy and matter that sustains them, sometimes altering it for their betterment and sometimes for their detriment.

The power of awareness to affect its own future outside the

directives innate in mechanical interactions creates responsibility. It creates the ability to choose paths leading to continuance or extinction, to a balanced or a troubled existence, to a measured coexistence with the mechanical Universe or to becoming a destructive force within it.

The difference between aware life as the result of innate patterns becoming emergent, when all possible arrangements of energy and matter are tested, or as the result of completely random occurrences, is of no consequence. Aware life exists, is rare, and is widely disbursed in the Universe and, like a rare element with useful qualities, is valuable.

In a Universe of energy and inorganic matter, organic matter comprises only a small percentage of the total mix, and of that small percentage, organic matter capable of replication, an even smaller percentage, and the self-awareness of Man an even smaller percentage.

Man carries an alternate universe in his collective synaptic reservoir. It is dependent on the random changes around him but is also capable of controlling some of the forces that dictate his future. The rarity of this aware state makes it, and all other aware states of awareness, valuable. The limited influence and small size of these alternate Universes makes them seem of no consequence, but their potential, if expanded and joined responsibly, makes them significant and this significance creates the same mandates for all beings with advanced states of awareness:

Explore and Learn
Be of good council
Be a good steward

57

TURNING THE UNIVERSE INSIDE OUT

Most of the reality around us escapes our attention. We assume things are solid because we can stand on them and they look solid, but when we investigate using scientific tools, things get strange. Matter isn't solid at all. It only feels and looks that way because of electro magnetic interactions between atoms. If we could see the space inside and between atoms, everything would disappear. Matter is almost entirely empty space.

The force that keeps us from floating off the floor also escapes our attention. We assume it is there because we feel it and can observe its effects. Things fall down, pressure increases the deeper we go in the ocean and decreases the higher we go in the atmosphere, all the effects of the invisible force called gravity.

We have evolved to observe and survive in our surroundings by interacting efficiently within the environment that has shaped us, but, by accident or plan, we have also acquired a curiosity, an enlarged brain and a complex language.

Using our advanced attributes, we have explored and conceptualized a universe well beyond that revealed by our senses and as we explore we create more questions; the most persistent being, what creates gravity?

Most of our progress in understanding gravity has resulted from individuals looking at the weakest of forces from new perspectives and asking new questions. Newton did this by defining matter as having mass and gravity as an attractive force between objects having mass. Einstein did this when he envisioned gravity, not as an attractive force, but as a geometric warping of space and time caused by objects having mass.

The puzzles associated with gravity involve both cosmology and atomic theory and both use the concept of mass as a comparative measure. Mass is referenced in atomic measurements, as electron volts and in the periodic table as atomic weights.

In the macro world mass is measured by the resistance an object presents to a change in velocity or by the pressures it creates when halted. The pressure you feel on your feet when standing is the combined mass attraction of the earth and your body. The pressure you feel in your inner ears when you ascend or descend in an airplane is the combined mass attraction of the air and the earth squeezing a compressible gas into a gradient density with the greatest pressure at the surface and the least at the top of the atmosphere.

The increasing pressure you feel when scuba diving, is the cumulative effect of an incompressible fluid being attracted toward the center of the earth with pressure increasing with depth. Gravity does all these things and more. It keeps planets and moons in orbit by matching their angular momentum. It creates compressive forces inside stars strong enough to fuse atoms and at its most extreme turns space inside out, and creates black holes.

Gravity gathers matter and compresses it creating differences in pressure and density. From these differences another force emerges, *Buoyancy*.

Buoyancy occurs in gravitational fields when an area of less pressure or density is surrounded by an area of greater pressure or density, as the situation that causes a hot air balloon rises or a submerged beach ball rushes to the surface when released under water, buoyancy acts in an opposite direction to gravity, a true anti gravity effect.

In un-accelerated zero gravity conditions, pressure/ density differences are not created by compression or weight and there are no anti gravity buoyant forces but, in un-accelerated zero gravity conditions, when small low density/pressure areas are surrounded by a large area of high density/pressure, the low density/pressure areas will tend to join. The joining force created by the pressure density difference pushes the low density/pressure areas together.

Think of space as a dense energy soup pushing outward, trying to reach a less energy dense state, (entropy).

Think of low energy dense areas in space as bubbles without a surface, fuzzy bubbles extending outward from a central point with their lowest energy density at the center. Unlike soap bubbles, these low energy dense areas have no membrane surface. They extend outward with their low energy effect dissipating as the square of the distance from their center, (much like gravity).

With their density effect dissipating with the distance from their center, low energy dense space bubbles can overlap and crowd together, until another force keeps their centers separated. As the bubbles accumulate their low energy density effect compounds and the difference between the high energy density of space and the joined low energy dense bubbles increases.

The tiny areas of low energy density I am referring to are the point energy particles that make up protons and neutrons, (quarks).

The high energy density of space I am referring to is dark energy, (energy with a density exceeding that of matter).

Dark matter, in the inside-out universe I am describing, is the result of large clouds of diffuse particles, possibly neutrinos, with a slightly lower energy/density than the energy field pervading the rest of space. The gathering power of low energy dense protons is the result of a buoyant energy force acting away from the high energy density of space.

From this perspective, visible matter is a low-energy dense state existing in a high-energy dense field. Gravity is the inward buoyant force created by a difference in energy state densities, and the buoyant force of gravity increases as negative pressure areas, (atoms), crowd together. With less room to move the vibrational rate of atoms is increased. We interpret this increase in vibrational rates as a raise in temperature and when we increase a temperature, by adding energy, we are lowering the energy density state of the atoms within a compound causing it to jump to a lower state, (solid to liquid, liquid to gas, and gas to plasma).

A black hole is the lowest state of energy/density possible, an extreme low-energy-dense state that collapses quark energy fields, transforms material states to a glass condensate without the ability to emit photons.

If space or matter has a grater energy density is to be determined. Weight, mass, inertia, pressure and black holes can be described with very similar equations. Thinking of space as having a greater energy density and matter less, helps explain things in cosmology, like the dominance of dark energy, and things in particle physics, like the relationship of quarks and the strong force to gravity.

To avoid confusion with older concepts the difference in density I am proposing is a difference in energy density, and the buoyancy we are substituting for gravity is energy buoyancy. Energy buoyancy operates like pressure buoyancy by seeking less dense areas, but energy buoyancy moves things toward material objects by creating spherical low energy dense areas in space ranging from atomic size areas surrounding individual protons to macroscopic stars and galaxies. Space-time is warped as it is in general relativity, but the mechanism is different, space is decompressed by matter instead of compressed, and gravity is a buoyant force instead of an attractive force

A reversed energy density condition is reasonable considering that the strong force, created by quarks, makes up 99% of the protons mass energy and has a negative energy component. It is made even more reasonable considering the recent discovery of dark energy. Within the inside-out universe I propose, matter exists in all its forms just as it did before, light travels as before and is slowed and diverted by areas affected by matter. Space, however becomes a denser energy medium with billions of small low-energy-dense areas interacting.

The concept of a buoyant energy dense state is easiest to understand as an isolated local condition. A proton interacts with the energy of space through the strong force that binds its quarks by disbursing some of the energy around it creating a dimple of lower energy density.

The proton remains less energy dense than its surroundings even in the reduced energy dense areas and remains fixed in the lowest energy dense position at the center of the dimple it created. As other protons with their own dimples come close they are drawn together, the dimples join and increase their intensity and size and draw more matter toward their centers in the same way a balloon moves toward an area of lower density.

As more protons and atoms join, the energy density at the center becomes even lower, high-energy space between the low energy dimples is reduced, and the influence of the growing low-density area expands. As smaller material groupings join forming larger groups they are assimilated and are drawn to the center, or if joining with angular momentum, may retain their own low energy density identity and orbit the center.

With matter and space now related,

- inertia becomes the adjustment of energy-density balances,
- black holes become the result of a compressive buoyancy causing matter to compress to a zero energy dense state,
- the anomalous rotation of galaxies is reconciled,
- The accelerating expansion of the universe, is partially explained by entropy, and partially by matter creating a concentration of higher energy density states around the reduced energy dense areas matter creates.
- Mass as a concept and as a measure of energy remains the same but with a negative exponent.

We have gone from thinking of space as having attributes similar to air and searching for its effect on the speed of light through an either, only to find that light traveled at the same speed in all directions regardless of any movement by an emitting object. We then assumed the universe was static only to find it was expanding. We then had our preconceptions altered again by the proposition that space was warped by matter and causes gravity. We then discovered that the universe is expanding at an accelerating rate and proposed a huge amount of invisible energy distributed throughout space as the cause.

What I am proposing is even more radical. I am proposing that we have been looking at the universe assuming space was empty, and that we, and everything around us, was solid, when in fact we, and all matter, are bubbles of low-energy density with space being the more dense energy state.

We have been looking at the universe with our evolved sensory organs with light our primary tool convinced that we understood the relationship between matter and energy, and as long as space remained a vacuum, our theories held true. Unfortunately, we have discovered that space is not empty, it is dominate, and light can't help us examine its nature.

We gave up our self-assumed and self-righteous belief that the universe was created especially for us, reluctantly. Now, to make sense of nature's latest revelations we need to acknowledge that we may be made of secondary stuff, looking at only a tiny bit of reality as reflected by light and, once again, rethink many of our assumptions

www.ingramcontent.com/pod-product-compliance
Lightning Source LLC
Chambersburg PA
CBHW030936180526
45163CB00002B/582